Sustained Simulation Performance 2019 and 2020

Michael M. Resch · Manuela Wossough ·
Wolfgang Bez · Erich Focht · Hiroaki Kobayashi
Editors

# Sustained Simulation Performance 2019 and 2020

## Proceedings of the Joint Workshop on Sustained Simulation Performance, University of Stuttgart (HLRS) and Tohoku University, 2019 and 2020

*Editors*
Michael M. Resch
High Performance Computing Center (HLRS)
University of Stuttgart
Stuttgart, Germany

Manuela Wossough
High Performance Computing Center (HLRS)
University of Stuttgart
Stuttgart, Germany

Wolfgang Bez
Europe GmbH
NEC High Performance Computing
Düsseldorf, Germany

Erich Focht
Europe GmbH
NEC High Performance Computing
Düsseldorf, Germany

Hiroaki Kobayashi
Cyberscience Center
Tohoku University
Sendai, Japan

ISBN 978-3-030-68051-0 ISBN 978-3-030-68049-7 (eBook)
https://doi.org/10.1007/978-3-030-68049-7

Mathematics Subject Classification: 65-XX, 65Exx, 65Fxx, 65Kxx, 68-XX, 68Mxx, 68Uxx, 68Wxx, 70-XX, 70Fxx, 70Gxx, 76-XX, 76Fxx, 76Mxx, 92-XX, 92Cxx



View of the velocity profile and contours around the cylindrical obstacle captured at an instant (Fig. 4 of the paper by Neda Ebrahimi Pour: 'Brinkman penalization and boundary layer in high-order Discontinuous Galerkin')

This Springer imprint is published by the registered company Springer Nature Switzerland AG
The registered company address is: Gewerbestrasse 11, 6330 Cham, Switzerland

# Preface

The Workshop on Sustained Simulation Performance (WSSP) has a tradition of more than 15 years and was only interrupted once when the earthquake of March 11 made it impossible to bring the Japanese colleagues to Stuttgart. Nevertheless, we celebrated the 30th workshop in autumn 2019 and were looking forward to a productive year 2020. All this changed with the COVID-19 pandemic that started in December 2019 in Wuhan/China and quickly spread across the world. Both, the event in spring and in autumn had to be cancelled as physical meetings were difficult if not impossible.

In this difficult situation, we decided not to throw away all the preparatory work for the two workshops but to publish the prepared talks and paper in our annual volume, which you are holding in your hands right now.

This volume brings together a series of papers devoted to the question of sustained simulation performance and related fields. All articles were part of the Workshop on Sustained Simulation Performance (WSSP) of October 2019 and of the planned workshop in March 2020 and authors were so generous to provide us with these works even though COVID-19 made it difficult to keep the schedules. We are hence rather grateful to all authors for their efforts in these difficult times.

High Performance Computing is facing a number of challenges and is looking at a number of opportunities. Among them are issues of performance and scalability but also new topics like the use of High Performance Computers in the field of Artificial Intelligence. At the horizon, we already see the advent of Quantum Computers. These issues are reflected in the papers presented.

In the first part of the book, we look at performance and power. Hiroaki Kobayashi and Kazuhiko Komatsu look into the NEC SX-Aurora TSUBASA system. José Miguel Montañana Aliaga, Alexey Cheptsov and Antonio Hervás address energy efficiency issues that increasingly shape the operation and usage of high performance computing systems and play a major role in optimizing for compute performance. Erich Focht looks at vector engine offloading a technology that has an impact on future performance.

In the second part about Numerics and Optimization Yujiro Takenaka, Mitsuo Yokokawa, Takashi Ishihara, Kazuhiko Komatsu, and Hiroaki Kobayashi look at the optimization of a DNS code on vector technology. Matthias Meinke, Ansgar Niemöller, Sohel Herff, and Wolfgang Schröder tackle the problem of load balancing for coupled simulation methods, a topic increasingly important as performance allows to simulate multi-physics applications. Finally, Neda Ebrahimi Pour, Nikhil Anand, Felix Bernhards, Harald Klimach and Sabine Roller look at high-order Discontinuous Galerkin.

Papers on Data Handling and New Concepts are collected in the third part. Patrick Vogler and Ulrich Rist look into large data handling and specifically into lossy compressors for CFD simulation data. Li Zhong has a look into a field that requires High Performance Computing but is so far hardly associated with it: Stream Data Analysis. Naweiluo Zhou brings the world of clouds in the field of High Performance Computing when looking at Containerization and Orchestration. Alexey Cheptsov carries High Performance Computing to geoscience applications based on CFD simulations using micro-services.

Finally, part four of this volume looks into trends that evolve in High Performance Computing and will shape the future of the field in the coming years. Michael Resch gives an overview of the role of Machine Learning and Artificial Intelligence while Denis Hoppe presents trends and emerging technologies in Artificial Intelligence.

We hope that the reader finds something interesting in this volume and express our hope that with the end of COVID-19 scientific life will go back to normal, allowing us to further intensive the scientific work and exchange in the field of Sustained Simulation Performance.

Stuttgart, Germany
November 2020

Michael M. Resch
Manuela Wossough

# Contents

**Performance and Power**

**Performance Evaluation of SX-Aurora TSUBASA and Its QA-Assisted Application Design** ... 3
Hiroaki Kobayashi and Kazuhiko Komatsu

**Towards Energy Efficient Computing Based on the Estimation of Energy Consumption** ... 21
José Miguel Montañana Aliaga, Alexey Cheptsov, and Antonio Hervás

**Speeding Up Vector Engine Offloading with AVEO** ... 35
Erich Focht

**Numerics and Optimization**

**Optimizations of DNS Codes for Turbulence on SX-Aurora TSUBASA** ... 51
Yujiro Takenaka, Mitsuo Yokokawa, Takashi Ishihara, Kazuhiko Komatsu, and Hiroaki Kobayashi

**Dynamic Load Balancing for Coupled Simulation Methods** ... 61
Matthias Meinke, Ansgar Niemöller, Sohel Herff, and Wolfgang Schröder

**Brinkman Penalization and Boundary Layer in High-Order Discontinuous Galerkin** ... 85
Neda Ebrahimi Pour, Nikhil Anand, Felix Bernhards, Harald Klimach, and Sabine Roller

**Data Handling and New Concepts**

**Handling Large Numerical Data-Sets: Viability of a Lossy Compressor for CFD-simulations** ... 97
Patrick Vogler and Ulrich Rist

**A Method for Stream Data Analysis** ... 111
Li Zhong

**CFD Simulation with Microservices for Geoscience Applications** ....... 121
Alexey Cheptsov

**Containerization and Orchestration on HPC Systems** ................. 133
Naweiluo Zhou

**Trends in HPC and AI**

**The Role of Machine Learning and Artificial Intelligence in High Performance Computing** ............................................. 151
Michael M. Resch and Bastian Koller

**Trends and Emerging Technologies in AI** ............................ 163
Dennis Hoppe

**Synergetic Build-up of National Competence Centres All over Europe** ........................................................ 183
Bastian Koller and Natalie Lewandowski

# Performance and Power

# Performance Evaluation of SX-Aurora TSUBASA and Its QA-Assisted Application Design

**Hiroaki Kobayashi and Kazuhiko Komatsu**

**Abstract** In this article, we present an overview of our on-going project entitled, *R&D of a Quantum-Annealing Assisted Next Generation HPC Infrastructure and its Applications*. We describes our system design concept of a new computing infrastructure toward the Post-Moore era by the integration of classical HPC engines and a quantum-annealing engine as a single system image and a realization of the ensemble of domain specific architectures. We also present the performance evaluation of SX-Aurora TSUBASA, which is the central system of this infrastructure, by using world well-known benchmark kernels. Here we discuss its sustained performance, power-efficiency, and scalability of vector engines of SX-Aurora TSUBASA by using HPL, Himeno and HPCG benchmarks. Moreover, As an example of the quantum annealing assisted application design, we present how a quantum annealing data processing mechanism is introduced into a large scale data-clustering.

## 1 Introduction

The vector computing technology, in which a single instruction can handle multiple and even many data simultaneously, is now fundamental for high performance computing platforms. Modern high performance microprocessors such as Intel Xeon, AMD, ARM, and NVIDIA GPU processors in addition to traditional NEC SX vector processors are continually enhancing their vector computing capability to boost the performance as shown in Fig. 1. However, to fully exploit the potential of the vector computing capability, memory subsystems that provide a large amount of data to a plenty of vector processing units play an important role. If the memory performance is not enough compared with the vector computing performance, the sustained performance in the execution of practical applications is limited due to the bottleneck of the memory subsystem.

H. Kobayashi (✉) · K. Komatsu
Tohoku University, Sendai 980-8579, Japan
e-mail: koba@tohoku.ac.jp

K. Komatsu
e-mail: komatsu@tohoku.ac.jp



M. M. Resch et al. (eds.), *Sustained Simulation Performance 2019 and 2020*,
https://doi.org/10.1007/978-3-030-68049-7_1

The NEC SX vector supercomputer series has been designed and developed based on the balanced architecture with a high B/F rate, a ratio of the memory performance in Bytes/s to the calculation performance in flop/s. Higher B/F ratios mean that the processor design is focused on increasing not only the peak performance but also the memory bandwidth to improve the sustained simulation performance. Figure 2 shows the trend in memory bandwidth of leading HPC processors. From the figure, we can observe that the latest SX vector processor SX-Aurora TSUBASA delivered in 2018 has 24x per core and 10x per socket higher memory bandwidth than those of Intel Xeon Skylake, even though Fig. 1 suggests that the difference in the peak performances of a core and a socket between the SX-Aurora TSUBASA and Xeon is not so significant, just 4x and 1.6x higher than Xeon, respectively.

However, when taking a look at the recent semiconductor technology for the fabrication of HPC processors, its development speed is gradually slowing due to the limitation in the miniaturization of transistors on the die and its related harmful effects such as heat dissipation and power consumption problems. This means that we are facing the end of Moore's Law. Under such a circumstance, post-Moore's information processing technologies that directly use physical effects and biological mechanisms such as Quantum computing, Brain-Inspired computing, etc. are drawing much attention as emerging ones to make a break-through in computing. Among them, quantum annealing that uses the quantum fluctuations to find the best combination of possible candidates under given conditions for solving so-called, Combinatorial optimization problem, is going into the commercial phase.

As many important problems such as traffic flow control, logistic control, AI training, etc. can be categorized into the class of the combinatorial optimization problem, quantum annealing is expected to become an accelerator to solve the combinatorial optimization problem. The D-Wave system, a Canadian startup company, has developed and commercialized the first transverse magnetic field type quantum annealing Chips and Systems in the world. By using the commercial systems, many research teams of Google, NASA, Volkswagen, Lockheed, Tohoku University, etc. get involved in R&D of emerging applications using quantum annealing especially that need solving combinatorial optimization problems, and successfully show its possibility that could outperform classical HPC systems toward the post-Moore era.

To satisfy high and never-ending demands and expectations for high-performance computing even after the end of Moore's law, we have been conducting R&D of a new generation high-performance computing infrastructure that incorporates quantum annealing into a vector computing platform since 2018 [1]. Even after the end of Moore's law, we believe that the silicon is a fundamental material just like concrete and steel for constructing buildings and bridges, and classical computers using matured materials are deployed everywhere as a social infrastructure. Of course, the computer architecture developments are strongly required ever before, and their contributions to the performance improvements are becoming larger than ever. Therefore, we are working together with NEC to explore the future of the vector computing technology under the consideration of current and emerging applications.

**Fig. 1** Trend in peak performance

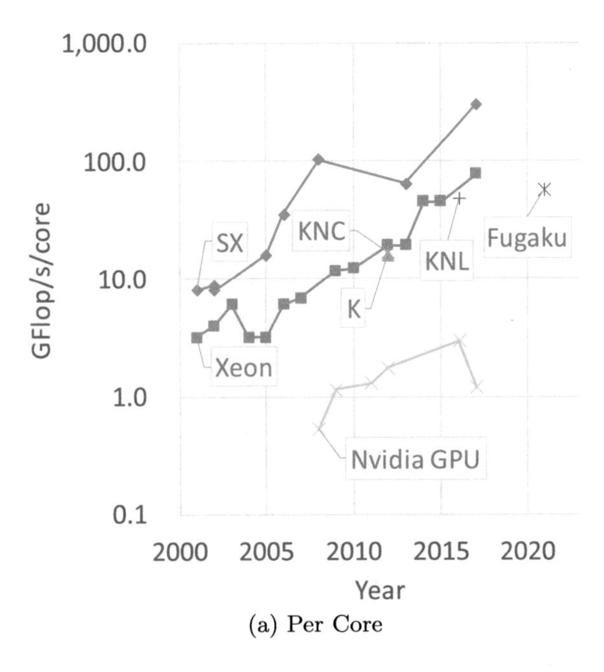


(a) Per Core

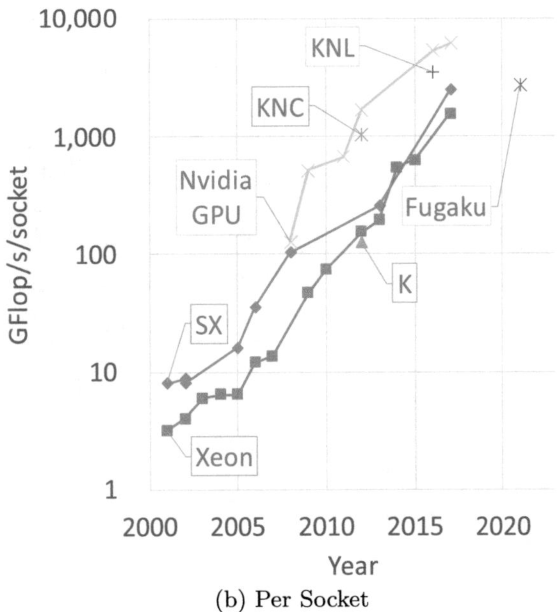


(b) Per Socket

**Fig. 2** Trend in peak memory bandwidth

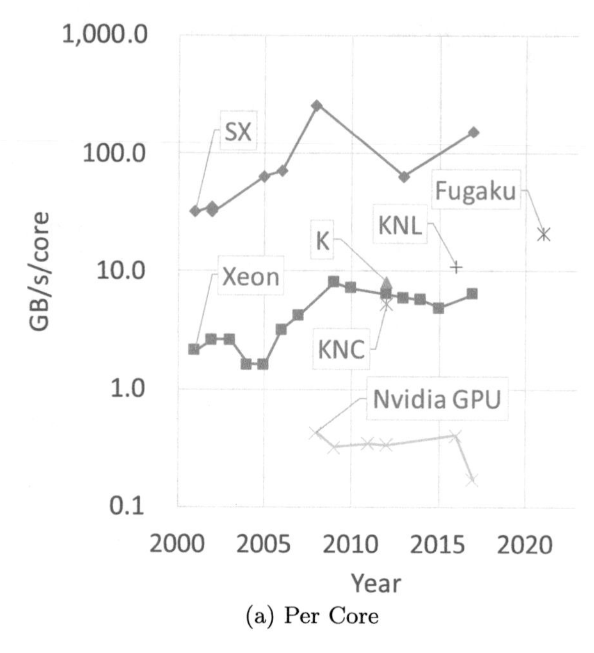


(a) Per Core

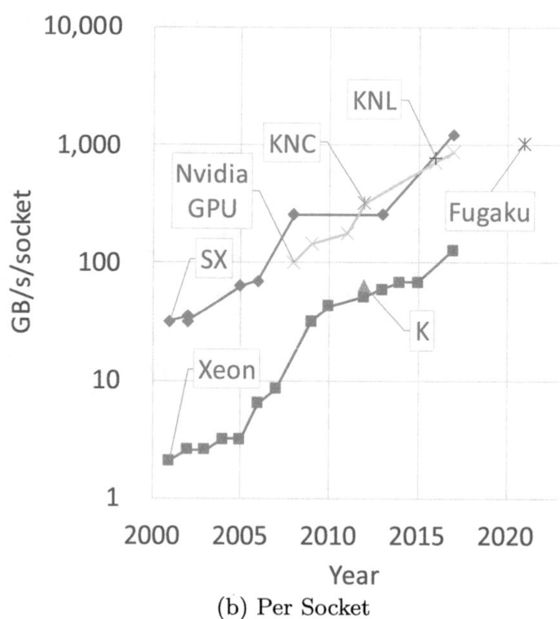


(b) Per Socket

Design of high-band memory subsystems is also an important issue and should be combined together with the vector computing architecture design by introducing advanced technology such as 3D stacking and silicon photonics. Here, we want to emphasize that the traditional general-purpose architecture design will not work anymore, especially for high-performance computing because the general-purpose functionality of computers leads to an inefficiency in computing and power-consumption across a wide variety of applications. Even in the case where one computing architecture shows a significant performance improvement in some applications, this architecture with a sophisticated mechanism to realize the general-purpose functionality may need a lot of power and silicon budgets to keep the performance, resulting in low power-efficiency. In addition, this architecture may not work well in the other applications because some general-purpose functionality does not contribute to its performance with the target applications. We think that the long lasting *spell* of "One fits all" in the computer design is no make sense in the Post-Moore era when considering the trade-off between performance and power&silicon budgets, and design, development, and deployment of architectures best suited for a specific domain of applications are mandatory to maximize the sustained performance as well as its power-efficiency.

To realize the general-purpose functionality of computing platforms in such a situation, orchestrating the different types of architectures for different domains of applications is feasible, and we name this new paradigm of computing architecture design approach to realization of the general-purpose functionality of platform is "*ensemble of domain-specific architectures*" approach. In the platform developed based on this concept, the whole applications or part of it is executed on appropriate individual architectures. The vector-computing equipped with a high-bandwidth memory subsystem is a domain-specific architecture for memory-intensive applications/kernels. If the high-performance vector-computing architecture with a moderate memory bandwidth subsystem may be considered a domain-specific architecture for computation-intensive applications/kernels. Besides, quantum annealing machines are classified into a domain-specific architecture for combinational optimization programs. In our R&D project, we try to realize a hybrid computing environment of a vector computing platform and a quantum annealer (QA) hosted by an X86 processor, in which kernels related to combinational optimization problems and memory-intensive kernels of an application are appropriately offloaded to QA and vector processors, respectively.

In this article, we present an overview of our on-going project entitled, *R&D of a Quantum-Annealing Assisted Next Generation HPC Infrastructure and its Applications*. Section 2 describes our system design concept of a new computing infrastructure toward the Post-Moore era by the integration of classical HPC engines and a quantum-annealing engine as a single system image and a realization of the ensemble of domain specific architectures. In Sect. 3, we present the performance evaluation of SX-Aurora TSUBASA, which is the central system of this infrastructure, by using the world well-known benchmark kernels. Here we discuss its sustained performance, power-efficiency, and scalability of vector engines of SX-Aurora TSUBASA by using HPL, Himeno, and HPCG benchmarks. Section 4 presents an example of the quantum annealing assisted application design. We discuss how a quantum annealing data processing mechanism is introduced into a large scale data-clustering. Section 5 summarizes the article.

## 2 Quantum-Annealing Assisted Next Generation HPC Infrastructure R&D

Figure 3 shows a stack representation of the target infrastructure to be developed in this project. At the lowest layer, hardware platforms are configured, and we place the D-wave machine as a quantum-annealing engine and NEC SX-Aurora TSUBASA as classical computing engines.

NEC SX-Aurora TSUBASA is the latest vector system and consists of one or more vector engines hosted by an X86 processor. Each vector engine is composed of a vector processor and a memory subsystem. On the vector processor, eight high performance cores are integrated, and are connected to the memory subsystem at a 1.22 TB/s and 1.53TB/s in the cases of the 10B and 20B models, respectively. Each core of the 20B model provides 307.2 Gflop/s for double-precision (DP) and 614.4 Gflop/s for single-precision (SP) floating-point calculations, and the 10B model has 2.15 Tflop/s and 4.3 Tflop/s for DP and SP operations, respectively. As a result, a VE processor with eight cores provides up to 2.45 Tflop/s (DP) and 4.91 Tflop/s (SP) floating-point performances in the case of the 20B model and 2.15 Tflop/s (DP) and 4.3 Tflop/s (SP) floating-point performances in the case of the 10B model. This high vector processing capability supported by a high memory bandwidth is expected to realize high-sustained performance in a wide variety of science and engineering applications, especially memory-intensive applications. At the same time, an X86 processor named Vector Host, which is equipped with a Xeon gold processor, is attached to vector engines. As the vector host provides the standard LINUX programming environment and OS functions, the LINUX system calls are automatically offloaded from the vector engine to the vector host in program execution as needed. As a result, SX-Aurora TSUBASA has a great potential of high performance vector

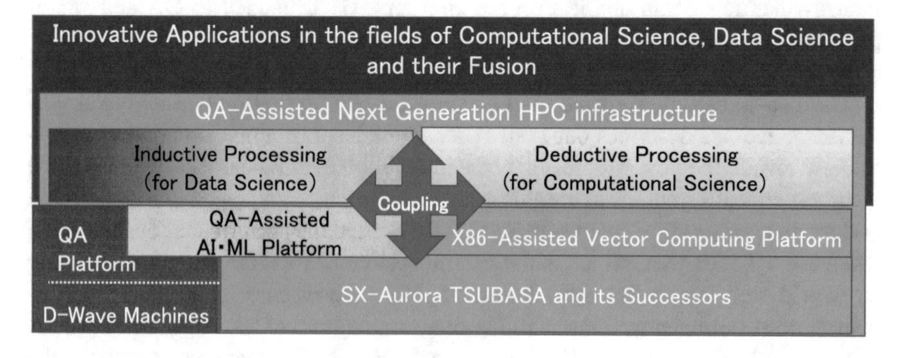


**Fig. 3** A stack representation of the target infrastructure

processing capability, while providing a standard programming environment. At the same layer, we introduce a D-Wave 2000Q as a quantum annealing engine. Program kernels for solving combinatorial problems could be offloaded to this engine through the programming interfaces implemented at the upper level of this layer.

Over the hardware engines layer, we construct two fundamental computing environments: a deductive computing environment and an inductive computing environment. The deductive computing environment is prepared to accelerate conventional simulations in the fields of computational science and engineering, and is highly-optimized for exploiting the potential of SX-Aurora TSUBASA. In this environment, effective vector computing-scalar computing hybrid is realized by heterogeneous computing of vector-engines and X86 processors (vector host). We also construct the inductive computing environment for AI-based data science applications at the same level. The inductive computing environment basically consists of application interfaces to the quantum-annealing engine for combinatorial problems and several standard AI-ML platforms such as Tensorflow and Frovedis (SparK compatible), where the latter one is highly-optimized for SX-Aurora TSUBASA. Moreover, as it is reported that the quantum-annealing works well in Boltzmann learning and clustering, we try to bring its potential to the AI-ML environment to provide the best choice to users based on their demands in the program development. In this project, we will intensively evaluate performances of quantum-annealing and classical computing as an accelerator for combinatorial problems and AI-ML applications, and figure out their best mix through the R&D of inductive computing and deductive computing integrated applications. Design and evaluation of a QA-assisted clustering application will be discussed in Sect. 4. Finally, over these environments, we construct a transparent interface to access these environments in a unified fashion, and build emerging simulation-AI-ML integrated applications, which will contribute to the realization of the Society 5.0, defined as a highly data-driven society by using the cyber-physical systems.

## 3 Performance Evaluation of SX-Aurora TSUBASA by Using Benchmark Programs

In this section, we examine the potential of SX-Aurora TSUBASA through the performance evaluation by using three benchmarks: HPL, HPCG, and Himeno in comparison with leading HPC processors and GPUs.

### 3.1 *HPL Performance*

HPL (High Performance LINPACK) is a computation-intensive benchmark used for TOP500 ranking. In this evaluation, we compare the sustained performance and its power consumption of SX-Aurora TSUBASA with those of AMD EPYC Naples and Rome. Specifications of evaluated systems are listed in Table 1.

**Table 1** Specifications of the evaluated systems

| CPU | AMD EPYC Naples 7601 | AMD EPYC Rome 7H12 | NEC SX-Aurora TSUBASA Type 20B |
|---|---|---|---|
| # of Cores/Socket | 32 | 64 | 8 |
| # of Sockets/Node | 2 | 2 | 1 |
| # of Cores/Node | 64 | 126 | 8 |
| Peak Perf/Node (Gflop/s) | 1,126 | 5,325 | 2,458 |
| Peak Perf/128 nodes (Tflop/s) | 144.1 | 681.5 | 314.5 |

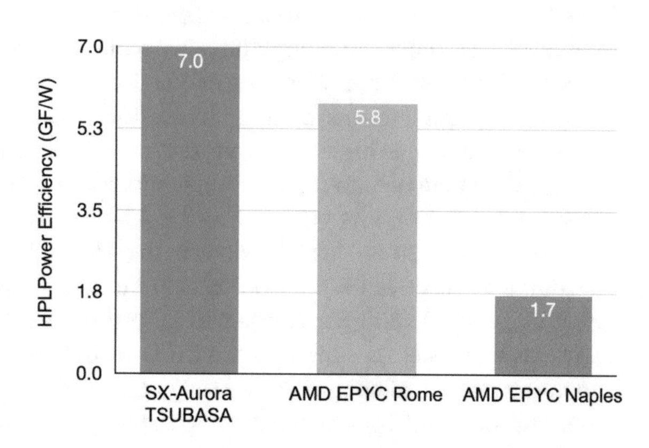


**Fig. 4** Power efficiency of HPC on 128 nodes

The evaluation results show that the HPL performance and power consumption of EPYC Naples, EPYC Rome, and SX-Aurora TSUBASA are 108.2 TF/64 kW, 477.1 TF/81.9 kW, and 267.7 TF/38.0 kW, respectively, when running at 128 nodes. Figure 4 shows the power efficiency (HPL effective performance/power consumption) for 128 nodes of parallel HPL execution. SX-Aurora TSUBASA is 4.1x more efficient than EPYC Naples and 1.2x more efficient than EPYC Rome for 128 nodes of parallel HPL execution. Generally, x86 processors are designed with more emphasis on the computational performance than the memory bandwidth. On the other hand, SX-Aurora TSUBSA is designed with more emphasis on the memory bandwidth than on the computational performance. However, the low power consumption design of the SX-Aurora TSUBASA shows that its power efficiency is superior to that of the X86 processor even in the case of HPL, which is considered the representative computation-intensive benchmark.

## 3.2 *Himeno Benchmark Performance*

The Himeno benchmark is an incompressible flow analysis code to evaluate the performance of a computer, and measures the processing speed of the main loops when solving Poisson's equation by Jacobi's iterative method. As a feature, it is a bench-

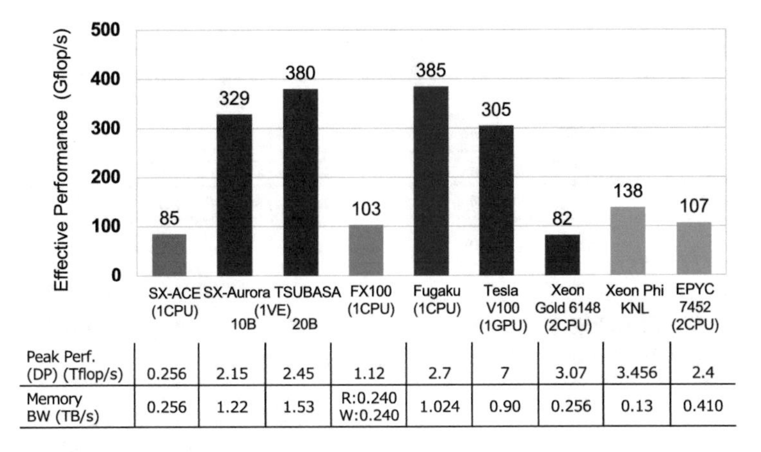


| | SX-ACE | SX-Aurora 10B | SX-Aurora 20B | FX100 | Fugaku | Tesla V100 | Xeon Gold 6148 | Xeon Phi KNL | EPYC 7452 |
|---|---|---|---|---|---|---|---|---|---|
| Peak Perf. (DP) (Tflop/s) | 0.256 | 2.15 | 2.45 | 1.12 | 2.7 | 7 | 3.07 | 3.456 | 2.4 |
| Memory BW (TB/s) | 0.256 | 1.22 | 1.53 | R:0.240 W:0.240 | 1.024 | 0.90 | 0.256 | 0.13 | 0.410 |

**Fig. 5** Sustained performance of the Himeno benchmark

mark that depends largely on the performance of the computer's memory bandwidth, mainly because of the high cost of 19-point stencil calculations. We evaluate the sustained performance of the Himeno benchmark on SX-Aurora TSUBASA, compared with those of modern HPC processors. In this evaluation, based on the MPI version of the Himeno benchmark, we optimize the computation domain partitioning for reducing the MPI communications as well as preferentially place reusable data in the LLC and maximize the vector length by loop unrolling.

Figure 5 shows the results of single-precision computation by the Himeno benchmark, which is the effective performance in flop/s. The horizontal axis indicates the system name. In addition to the performance of the two SX-Aurora TSUBASA systems (10B and 20B, each measured at 1VE), the performance of SX-ACE, FX100, Fugaku, Tesla V100, and EPYC 7452, each of which is measured on one processor, is shown here for comparison, where Fugaku's performance is based on the ISC2020 publication. Xeon Skylake measured with two processors are also shown. The lower part of the figure shows the system specs (Tflop/s value of double precision arithmetic performance and memory bandwidth performance in TeraBytes/s). The data is XL ($1024 \times 512 \times 512$), which has the largest problem size.

As the Himeno benchmark has a high memory load, the figure clearly suggests that the high memory bandwidth and high vector core and socket performance lead to a high effective performance in the Himeno benchmark. As a result of further optimization of the MPI version, we can achieve a 6% performance improvement on SX-Aurora TSUBASA 10B compared with that reported at SC18 [2].

Table 2 shows the B/F rate of each system and the efficiency of the Himeno benchmark based on the values shown in Fig. 5. Here, the efficiency is defined as a ratio of the effective performance of the Himeno benchmark to the peak performance of each system. A higher values mean an effective use of installed computation units. As the table shows, higher B/F rates lead to higher efficiencies. Therefore, we can confirm that a high B/F rate in addition to a high peak performance, e.g.. SX-Aurora

**Table 2** B/F and the Himeno efficiency of each system

| System | SX-ACE | SX-Aurora | | FX100 | Fugaku | Tesla V100 | Xeon Skylake | Xeon Phi | EPYC Rome |
|---|---|---|---|---|---|---|---|---|---|
| | | 10B | 20B | | | | | | |
| B/F | 1.00 | 0.57 | 0.62 | 0.43 | 0.38 | 0.13 | 0.08 | 0.04 | 0.17 |
| Efficiency (%) | 33.2 | 15.3 | 15.5 | 9.2 | 14.3 | 4.4 | 2.7 | 4.0 | 4.5 |

TSUBASA 20B in Fig. 5 and Table 2, is an important factor to extract the potential of processors, in particular, for memory-intensive applications.

## 3.3 HPCG Benchmark Performance

The HPCG benchmark measures the performance of a computer by solving the conjugate gradient method (CG method) with preprocessing using the Multi-Grid method for solving a simultaneous linear equation Ax = b with symmetric sparse matrices discretized by the finite element method. In the implementation of the HPCG benchmark on SX-Aurora TSUBASA, vectorization and parallelization of the hyperplane method are applied as its tuning. In the vectorization, we increase the sizes of loops in both the y and z directions as long as possible to maximize the vector length. Moreover, to reduce the access conflicts on the last-level cache of the vector processor, we use the partition mode that logically partitions on-chip vector cores into two groups, four each. By using the partition mode, the localization of LLC accesses within the group is enforced. These optimizations allowed us to achieve a sustained performance of 125.9 Gflop/s in the case of single VE, resulting in a high execution efficiency of 5.9%. For reference, according to the HPCG benchmark website (https://www.hpcg-benchmark.org), many of modern processors are in the 1–2% range. As the HPCG benchmark is memory-intensive, SX-Aurora TSUBASA shows a significant efficiency thanks to the balanced architecture with a high B/F rate.

Here, using SX-Aurora TSUBASA with 8 VEs of the 10B model, we tested the effects of logically dividing the inside of the vector processor into two groups and distributing the process as much as possible to multiple groups/VEs on the sustained performance. The partitioning mode splits the eight cores of a single VE socket into two logical groups to increase the locality of each LLC access, while the normal mode is the mode without partitioning. In addition, mode *compact* prioritizes the placement of processes into a single socket or logically partitioned cores. This means that after placement on one socket, or on all cores grouped by partition, processes are placed on the next VE socket or logical group.

Figure 6 shows the scalability of the HPCG benchmark regarding the effective performance and power efficiency when changing the number of VEs from 1 (8 cores) to 8 (64 cores). The horizontal axis shows the number of processes, which is

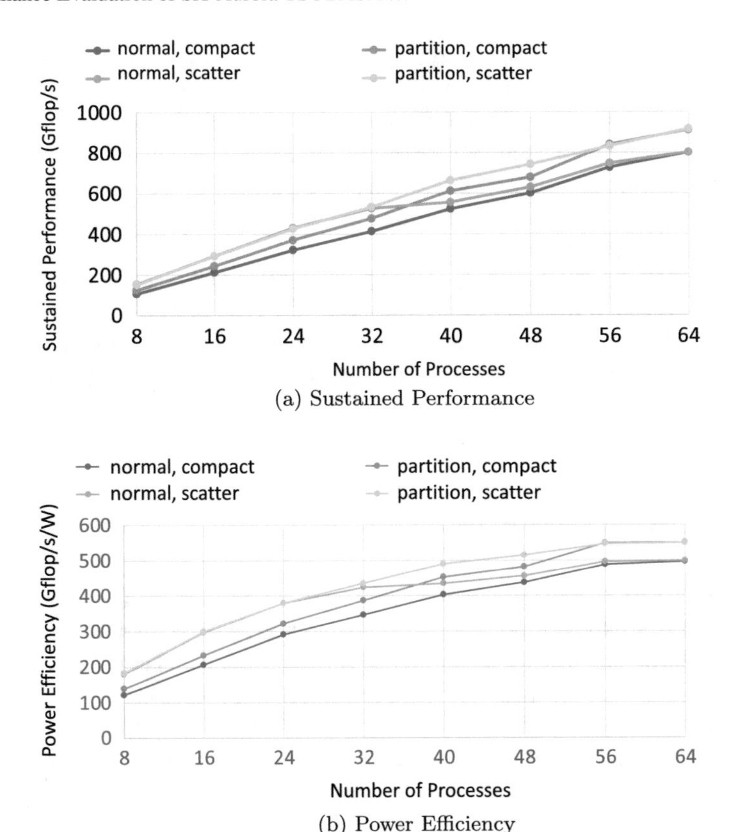


**Fig. 6** Trend in peak performance

equivalent to the number of cores. The vertical axis shows the effective performance and the power efficiency, where all four combinations of the normal/partition modes and the compact/scatter modes are examined. As it can be seen from the figure, the performance of combination (partitioning, scatter), which localizes the LCC accesses within the grope/socket and uses the scatter mode from the perspective of load balancing, leads to a stable and high scalability.

## 4 Data Clustering Using Quantum Annealing

Under the circumstance that the post-Moore era will come in the near future, many algorithms that utilize combinatorial optimization problems have been developed. For example, clustering algorithms, one of the fundamental algorithms in the field of data sciences and computer sciences, using combinatorial optimization problems have been recently developed [3–6]. In our R&D project, important target applications, such as analysis of time-series of data obtained from a real turbine and its digital twin and analysis of new material designs, require a clustering algorithm.

However, it is well known that the conventional clustering algorithm exponentially increases the computational cost as the numbers of data and features increase. Therefore, quantum annealing, which can solve a combinatorial optimization problem that minimizes an objective function under various constraints, is expected to suppress the computational cost.

This section describes a clustering algorithm using QA that is one of the information processing devices in the post-Moore era. This section especially focuses on QA-assisted agglomerative hierarchical clustering that uses combinatorial optimization problems and discusses the performance in terms of execution time and quality.

## *4.1 Clustering Algorithm*

Clustering is mainly divided into two ways; non-hierarchical clustering and hierarchical clustering. Non-hierarchical clustering is used when the number of clusters is known while hierarchical clustering is used when the number of clusters is unknown. In the cases that there are many problems in which the appropriate number of clusters is unknown, hierarchical clustering is effective. Since clustering results are hierarchically obtained, it is possible to select an appropriate number of clusters from the results after the clustering. Thus, it is not necessary to know the number of clusters in advance.

Although the execution time of hierarchical clustering is longer than that of non-hierarchical clustering, hierarchical clustering has a possibility to be accelerated by quantum annealing. An agglomerative hierarchical clustering, one of the hierarchical clustering methods, determines representative points of clusters by solving combinatorial optimization problems [3].

## *4.2 Quantum Annealing*

The number of combinations among multiple variables exponentially increases as the number of variables increases. Thus, it is not realistic to search for all combinations to find the best solution. Furthermore, it is difficult to find the best solution by the greedy search method that searches for combinations of small values of an objective function because the neighborhood search method easily encounters a local optimal solution.

To escape from the local optimal solution, simulated annealing and quantum annealing algorithms are often utilized. In simulated annealing and quantum annealing, even a non-optimal direction search that increases a value of an objective function is performed with a certain probability called a *fluctuation effect*. As the search progresses, the fluctuation effect is controlled not to occur often. Thus, it becomes possible to escape from the local optimal solution at the initial stage of the search, and then, the search only for the optimum direction is performed at the final stage of the search.

Quantum annealing is executed to search for solutions by *quantum fluctuations*. The smallest unit is called a *qubit*. One qubit can represent both "0" and "1" states simultaneously. The qubit in this superposition state becomes either "0" or "1" when observed. This indicates that the conventional representation with $2^N$ bits can be represented with $N$ qubits. Due to this phenomenon called the *superposition effect*, multiple combinations can be searched at the same time. As a result, the optimal solution can be quickly obtained.

In addition, even when a local optimal solution is encountered, it is possible to escape from the local optimal solution by the *quantum tunnel effect* that can go through a barrier.

In order to use quantum annealing, dedicated processors such as QA are necessary. Currently, D-Wave provides a quantum annealing processor, D-Wave 2000Q, as a commercial product. D-Wave 2000Q is equipped with 2048 qubits for quantum annealing. However, all qubits are not fully connected because the chimera graph structure is used for the connection among qubits. Therefore, only 64 qubits can be used to solve a problem that requires fully connected qubits.

To solve combinatorial optimization problems by D-Wave 2000Q, an objective function is formulated as an upper triangular matrix called *QUBO (Quadratic Unconstrained Binary Optimization)*. QUBO is obtained by transforming a Hamiltonian from an objective function.

## 4.3 Agglomerative Hierarchical Clustering Using QA

In order to perform agglomerative hierarchical clustering using QA, combinatorial optimization problems are offloaded to QA while the other computations are performed on the conventional computing systems. To offload to QA, a problem size is divided until QA can handle it. Since hierarchical clustering divides a large problem into small problems, it is very compatible with QA that has constraints of the number of qubits.

Figure 7 shows an overview of the agglomerative hierarchical clustering using QA. The agglomerative hierarchical clustering mainly comprises three steps: data partition, chunk coarsening, and chunk collapsing. By repeating these steps, the hierarchical structure of clustering is obtained.

In the data partition step, the entire data is repeatedly divided into two subsets called *chunks* in the maximum distribution direction. The maximum number of data of each chunk is decided by the size of QUBO that QA can solve. The data in Fig. 7a is divided into chunks, as shown in Fig. 7b.

In the chunk coarsening step, the representative point of each cluster is determined within a chunk. When there is $P = \{p^{(1)}, \ldots, p^{(n)}\}$ data in a chunk, a parameter $\epsilon$

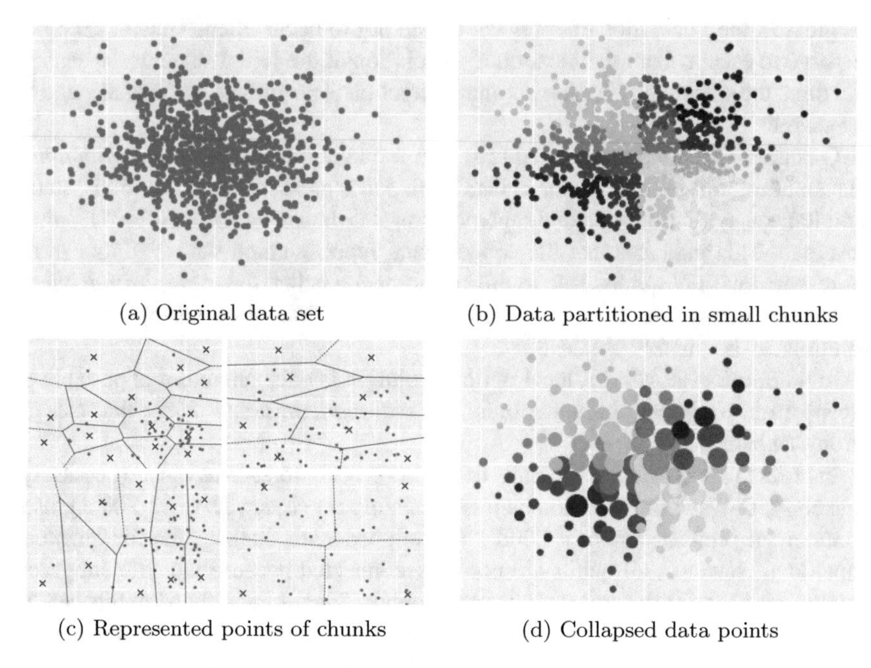


**Fig. 7** Overview of agglomerative hierarchical clustering

is introduced to measure the similarity between the data. In addition, the weighted graph $G^\epsilon = (P, E^\epsilon, w_P)$ with the vertex set $P$ and the edge set $E^\epsilon$ is considered. $E^\epsilon$ indicates a set of edges given by vertices that satisfies the distance $d(p^{(i)}, p^{(j)})$ is smaller than $\epsilon$. The weight $w_P(p^{(i)})$ of a point $p^{(i)}$ is defined as $w_i$. When a binary variable $s_i \in \{0, 1\}$ is given to $p^{(i)}$, $s_i$ indicates whether $p_{(i)}$ becomes the representative point or not. In addition, similarity matrix $N^{(\epsilon)}$ for $G^\epsilon = (P, E^\epsilon, w_P)$ is defined. When $d(p^{(i)}, p^{(j)}) < \epsilon$ is held, $N^{(\epsilon)}_{ij}$ becomes 1. Otherwise, $N^{(\epsilon)}_{ij}$ becomes 0.

By transforming the process of determining the representative points of a cluster into a maximum weighted independent set problem (MWIS), QA can be used. MWIS can be expressed in the following equation as a quadratic constrained quadratic program.

$$\begin{aligned} \underset{s\in\{0,1\}^n}{\text{maximize}} \quad & \sum_{i=1}^{n} s_i w_i \\ \text{subject to} \quad & \sum_{i=1}^{n}\sum_{i<j} s_i N^{(\epsilon)}_{ij} s_j = 0. \end{aligned} \tag{1}$$

The solution $s$ is the maximum independent set of the given weighted graph $G^\epsilon$. Therefore, all points $p^{(i)}$ that satisfy $s_i = 1$ can be treated as representative points. Equation (1) is an NP-hard MWIS problem for $G^\epsilon$. When the problem size becomes large, it becomes difficult to find an exact solution by the full search algorithm. In

QA, Eq. (1) is converted to QUBO. As shown in Fig. 7c, the representative points of each chunk are decided.

In the chunk collapsing step, the weights of other points in the same chunk are collapsed in the representative point. This can significantly reduce the number of data required for clustering. The final clustering result is obtained by repeating the data partition step, the chunk coarsening step, and the chunk collapsing step with a new data set.

## 4.4 Evaluation of Agglomerative Hierarchical Clustering Using QA

For evaluations, D-Wave 2000Q is used through the cloud service Leap. The annealing time is set to be 10 and 20 μsec. For the data partition step and the chunk collapsing step, Intel Xeon Gold 6126 is used.

For comparisons, the Greedy algorithm and SA are used for the chunk coarsening step as well as QA. For SA, OpenJij library version 0.0.9 [7] is used. The number of sweeps is set to be 500 or 1000.

Furthermore, the conventional clustering methods using various platforms such as CPU, NVIDIA Tesla V100, NEC Vector Engine(VE) Type 10B [8, 9] are evaluated. The clustering methods implemented on scikit-learn [10] such as k-means, Spectral Clustering, Agglomerative hierarchical clustering, DBSCAN, Birch, and mini-batch k-means are used. For GPU, k-means and DBSCAN implemented on Rapidsai version 0.12.0a [11] are used. For VE, k-means and Spectral clustering implemented on Frovedis version 0.9.5 [12] are used.

The Iris data set [13] is used for the evaluation of clustering. The data set has 150 data and four features. Since the optimal number of clusters in the data set is three, the execution time until the number of clusters reached three is measured in hierarchical clustering. For non-hierarchical clustering, the execution time when the number of clusters is set to three is measured.

Figure 8 shows the execution time. The x-axis shows the clustering methods and platforms. The y-axis shows the execution time. In the case of QA and SA, the chunk coarsening step is divided into the generation of QUBO, called *coarsening1*, and the solving QUBO by annealing, called *coarsening2*. Figure 8 shows that the execution times of coarsening in QA are shorter than those in SA and greedy. This is because the execution time of coarsening2 is short as QA could solve MWIS quickly. However, the execution times of collapsing in QA are longer than those in SA and greedy. This is because QA can find a larger independent set, causing an increase in the cost of the chunk collapsing. As the chunk collapsing is independent among different chunks, it can be accelerated by the parallelization.

Figure 8 also shows that the execution times of QA are longer than those of the conventional clustering methods by scikit-learn. This is because the conventional

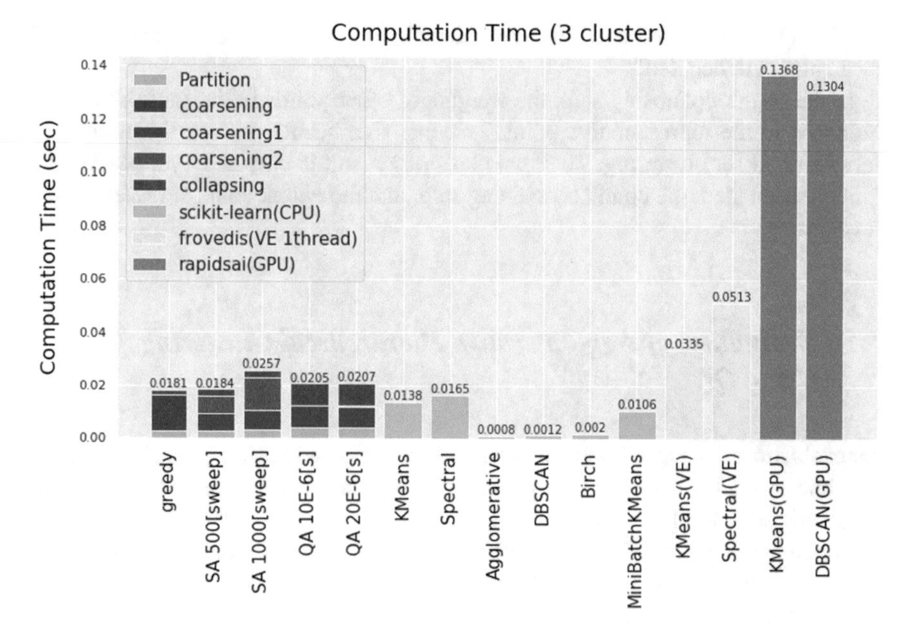


**Fig. 8** Execution times of various clustering methods

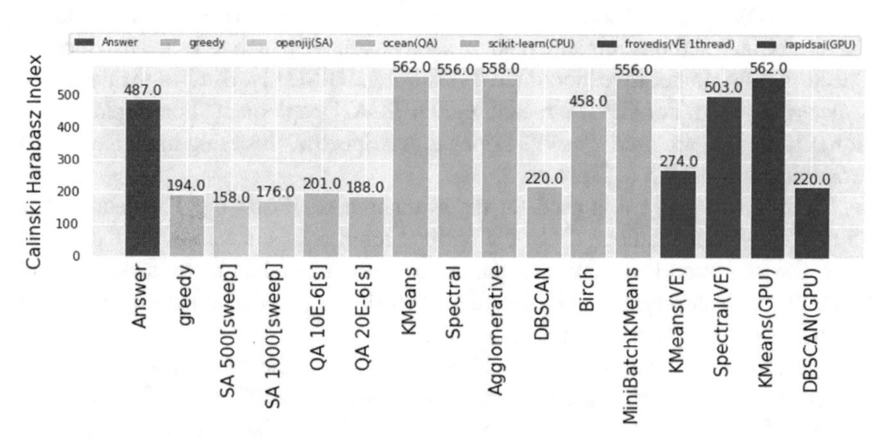


**Fig. 9** Qualities of various clustering methods

clustering methods in scikit-learn are well optimized. To accelerate the clustering on QA, the optimization of the Python implementation is necessary.

Furthermore, Fig. 8 also shows that the execution times of VE and GPU are longer than those of the others. This is because the problem size of the Iris data set is too small for VE and GPU. It is necessary to evaluate a larger data set for fair comparisons.

Figure 9 shows the quality of the clustering methods. The x-axis shows the clustering methods. The y-axis shows the Calinski-Harabasz(CH) index defined by the ratio of the sum of inter-cluster dispersion and intra-cluster dispersion [14]. The higher the index, the better the quality. The CH index of the correct classification result in the Iris data set is shown as *Answer*. The CH index of QA 10E-6[s] is higher than those of QA 20E-6[s], SA, and greedy. This is because QA of 10 μsec obtains a solution closer to the optimal solution of MWIS compared with QA of 20 μsec, SA, and greedy. Since many representative points are obtained, clustering could be performed without losing the characteristics of the original data, resulting in the improvement of the quality.

However, the CH indexes of the agglomerative clustering on QA, SA, and greedy are lower than those of scikit-learn, Frovedis, and Rapidsai. This is because the data partition of the agglomerative clustering affects the accuracy of MWIS in addition to the differences between the hierarchical clustering and the non-hierarchical clustering.

As shown in Fig. 9, some CH indexes are higher than that of Answer. Since the CH index is a value from the viewpoint of variance, the clustering algorithm may show higher quality than the correct classification result.

From these evaluations, it is clarified that QA has a possibility to accelerate the clustering since the coarsening step that can be offloaded to QA is much faster than the others. Currently, as there are some constraints such as the number of qubits, QA becomes one of the information processing devices in the post-Moore era. Moreover, simulated annealing for SX-Aurora TSUBASA has been being developed. Since there are no constraints such as the number of bits, it would be one of the candidates to perform simulated annealing.

## 5 Summary

This article describes our on-going project for design and development of a future HPC infrastructure, which incorporates quantum-annealing as a post-Moore information processing mechanism into the classical high performance computing infrastructure. Our design is based on the latest vector supercomputer system SX-Aurora TSUBASA and its successors, but for the specific kernels to handle combinatorial problems, they are offloaded to a quantum annealing machine in a transparent fashion from the user point of view. Even though the vector supercomputer system SX-Aurora TSUBASA has a great potential to accelerate conventional memory-intensive applications, it is shown that QA also has a possibility to accelerate data clustering by offloading combinatorial optimization problems. Therefore, the hybrid computing of the vector supercomputer system and the post-Moore system has a great potential to achieve further high-performance and high-sustained computing.

**Acknowledgements** Many colleagues get involved in this project, and great thanks go to faculty members of the Tohoku-NEC Joint Lab. at Cyberscience Center of Tohoku University. This project is supported by the MEXT Next Generation High-Performance Computing Infrastructures and Applications R&D Program.

## References


1. Resch, M., Kovalenko, Y., Bez, W., Focht, E., Kobayashi, H. (eds.): Sustained Simulation Performance 2018 and 2019. Springer International Publishing (2020)
2. Komatsu, K., Momose, S., Isobe, Y., Watanabe, O., Musa, A., Yokokawa, M., Aoyama, T., Sato, M., Kobayashi, H.: Performance evaluation of a vector supercomputer sx-aurora tsubasa. In: SC18: International Conference for High Performance Computing, Networking, Storage and Analysis, pp. 685–696. IEEE (2018)
3. Jaschek, T., Bucyk, M., Oberoi, J.S.: A quantum annealing-based approach to extreme clustering. In: Arai, K., Kapoor, S., Bhatia, R. (eds.) Advances in Information and Communication, pp. 169–189. Springer International Publishing, Cham (2020)
4. Kumar, V., Bass, G., Tomlin, C., Dulny, J.: Quantum annealing for combinatorial clustering. Quan. Inf. Process. **17**(2), 39 (2018)
5. Hastie, T., Tibshirani, R., Friedman, J.: The Elements of Statistical Learning: Data Mining, Inference and Prediction, 2nd edn. Springer (2009)
6. Kurihara, K., Tanaka, S., Miyashita, S.: Quantum annealing for clustering. In: Proceedings of the 25th Conference on Uncertainty in Artificial Intelligence, UAI 2009, pp. 321–328, 2009. 25th Conference on Uncertainty in Artificial Intelligence, UAI 2009; Conference date: 18-06-2009 Through 21-06-2009
7. OpenJij: Framework for the Ising model and QUBO
8. Yamada, Y., Momose, S.: Vector engine processor of nec's brand-new supercomputer sx-aurora tsubasa. In: Proceedings of a Symposium on High Performance Chips, Hot Chips, vol. 30, pp. 19–21 (2018)
9. Komatsu, K., Kobayashi, H.: Performance evaluation of SX-Aurora TSUBASA by using benchmark programs. In: Resch, M.M., Kovalenko, Y., Bez, W., Focht, E., Kobayashi, H. (eds.) Sustained Simulation Performance 2018 and 2019, pp. 69–77. Springer International Publishing, Cham (2020)
10. Pedregosa, F., Varoquaux, G., Gramfort, A., Michel, V., Thirion, B., Grisel, O., Blondel, M., Prettenhofer, P., Weiss, R., Dubourg, V., Vanderplas, J., Passos, A., Cournapeau, D., Brucher, M., Perrot, M., Duchesnay, E.: Scikit-learn: machine learning in Python. J. Mach. Learn. Res. **12**, 2825–2830 (2011)
11. RAPIDS Development Team. RAPIDS: Collection of Libraries for End to End GPU Data Science (2018)
12. Frovedis: Framework Of VEctorized and DIStributed data analytics
13. Dua, D., Graff, C.: UCI machine learning repository (2017)
14. Caliński, T., Harabasz, J.: A dendrite method for cluster analysis. Comm. Stat.-Theory Methods **3**(1), 1–27 (1974)

# Towards Energy Efficient Computing Based on the Estimation of Energy Consumption

**José Miguel Montañana Aliaga, Alexey Cheptsov, and Antonio Hervás**

**Abstract** The amount of computation power in the world keeps increasing as well as the computation needs by the industry and society. That increases also the total energy consumption on ICT, which reached the level of billions of dollars spent every year, as well as an equivalent emission print of millions of tons of $CO_2$ per year. That economical and ecological costs motivate us to search for more efficient computation. In addition, one more need for an efficient computer is the target of exascale computing and higher levels after that. We consider that it is needed a shift from considering only computation time when optimizing code, to also consider more efficient use of energy. To achieve energy-efficient computing, we consider that the first step considers recording the energy consumption of the algorithms used, and then using those results to select a more efficient energy algorithm among those available, which may require to increase the parallelization level and/or computation time, while still fulfill with the application requirements. Notice that cooling systems in the HPC may require to consume the same amount of energy as that consumed in the computing nodes, which means that the reduction of energy consumption due to efficient energy programming will also be doubled.

J. M. Montañana Aliaga (✉) · A. Cheptsov
High Performance Computing Center Stuttgart (HLRS), University of Stuttgart, Nobelstraße 19, 70569 Stuttgart, Germany
e-mail: jmmontanana@gmail.com

A. Cheptsov
e-mail: cheptsov@hlrs.de

A. Hervás
Instituto Universitario de Matemática Multidisciplinar, Universitat Politècnica de València, Camino de Vera s/n, 46022 Valencia, Spain
e-mail: ahervas@mat.upv.es

 
M. M. Resch et al. (eds.), *Sustained Simulation Performance 2019 and 2020*,
https://doi.org/10.1007/978-3-030-68049-7_2

# 1 Motivation

The energy consumption in the ICTs already represented a considerable amount of energy as well as its footprint in the emission of $CO_2$ in the past decades, as shown in the Table 1.

Moreover, previous studies show that ICT energy consumption increases every year, although the efficiency of ICT equipment increases [11].

Additionally, other studies like [13] shows that the energy consumption of data centers worldwide between 2010 and 2017 shown an accelerated increase. That shows in contrast with more conservatives predictions in 2010 which had proved to be wrong. Those old modes failed because did not consider the increase of energy consumption on computing nodes increases, as it happens for instance with the increase of the amount of RAM and architectural changes on the CPUs.

Current models, such as Andrae [4] and Belkhir [5] forecast the worst case as an exponential increase in the total energy consumption worldwide due to the Data Centers. Such worst-case seems not acceptable with the current ecological dispositions and governmental objectives, such European Commission 2015 have a more optimistic target of efficiency gains according to Moore's Law continues until 2030

**Table 1** Some reference estimations of energy consumption

| **Estimations of energy consumption of data centers** |
|---|
| U.S. data centers energy consumption: |
| In 2006 almost 61 TWH, which is about 1.5% of the total energy consumption in USA [6, 27] |
| In 2011 more than 100 TWH [2, 27] |
| Worldwide data centres use an estimated energy: |
| In 2005 was 150 TWh [16] |
| In 2019 was 200 TWh [20] |
| 100 TWH/year × 0.12 $/Wh = **12 billion USD$ in 2011 in the USA** |
| 100 TWH/year × 0.62 kg $CO_2$ = **62 million tones $CO_2$ in 2011 in the USA** |
| **Estimations of ICT worldwide energy consumption** |
| In 2007 was 710 TWh [19] |
| In 2012 was 920 TWh [8, 12] |
| In 2016 was 1059 TWh [24] |
| 920 TWh/year × 0.12 $/Wh = **110 billion USD$ per year** |
| 920 TWh/year × 0.62 kg $CO_2$ = **570 million tones $CO_2$ per year** |
| **Estimations of energy consumption of HPC centers** |
| Some reference values on the top supercomputers in Top500 list: |
| In 2011 Tianhe-1A 2.57 petaFLOPS 4.04 MW [23] |
| In 2013 Tianhe-2 consumes 17.6 MW + 6.4 MW cooling for 33.86 petaFLOPS [14] |
| In 2020 Fugaku consumes 28.33 MW for 0.415 exaflops [1] |
| 28 MWh/year × 0.12 $/Wh = **3.36 million USD$ per year** |
| 28 MWh/year × 0.62 kg $CO_2$ = **17 thousand tones $CO_2$ per year** |

[33]. It seems that those optimistic forecasts are unrealistic [17], and the common consensus among all parties is that efforts on efficient computing are needed.

On the other hand, the committee directed by the US DOE (Department of Energy) wrote a list of ten challenges needed to be solved needed to reach the Exascale computing, because "*an evolutionary approach to achieving exascale would not be adequate*" [18]. The first challenge is to achieve energy-efficient to "*achieve exascale using 20 MW of power*" [18].

Therefore, the targets in all the ICT scenarios require to reach energy-efficient computing, which can be achieved using more energy-efficient hardware and software that currently being used. The experts' report exposes "Without much more energy-efficient circuits, architecture, power conversion, power delivery, and cooling technologies, the total cost of ownership for exascale systems could be 10 times higher than today" [18].

In line with efforts at efficiency at the hardware level, we can mention that recently, in June 2020, Japanese supercomputer *Fugaku* [26] become the fastest in the world, reaching in a Top500 HPL with the result of 0.415 exaFLOPs, which is almost half exascale level. The design of the supercomputer is based on the use of more energy-efficient hardware, among which we can highlight the use of ARM processors, considered the most energy-efficient [25]. In particular, the supercomputer uses Fujitsu's 48-core A64FX CPUs. However, multiple processors working in parallel are needed to achieve the same computation time as a more powerful processor consuming fewer power [22], which means that software developers will need to increase the level of parallelism on their applications.

In this paper, we focus on providing a tool to reach a more efficient computation at the software level, it can help to save energetical and ecological costs, but also to improve energy-efficient computing needed for the exascale computing and higher levels after that.

## 2 Objectives

The data shown in the previous section on global economic spending and environmental impact motivates efforts to improve ICT efficiency since even a small percent improvement represents a significant amount of money as well as the $CO_2$ environmental footprint.

These improvements, clearly of interest to the industry and to society, are being the subject of multiple government-funded works [3, 7, 9, 28], among them we wish to highlight the ICT-Energy [15] project whose objective is efficient computing at all levels from small computing devices in systems distributed to large supercomputing centers. Among the proposals, they showed in their presentation at the University of York on 2016, the classification of energy efficiency at the software level was exposed so that governments and industry have a reference for the energy cost of different software solutions existing and thus have one more criterion for decision-making (Fig. 1).

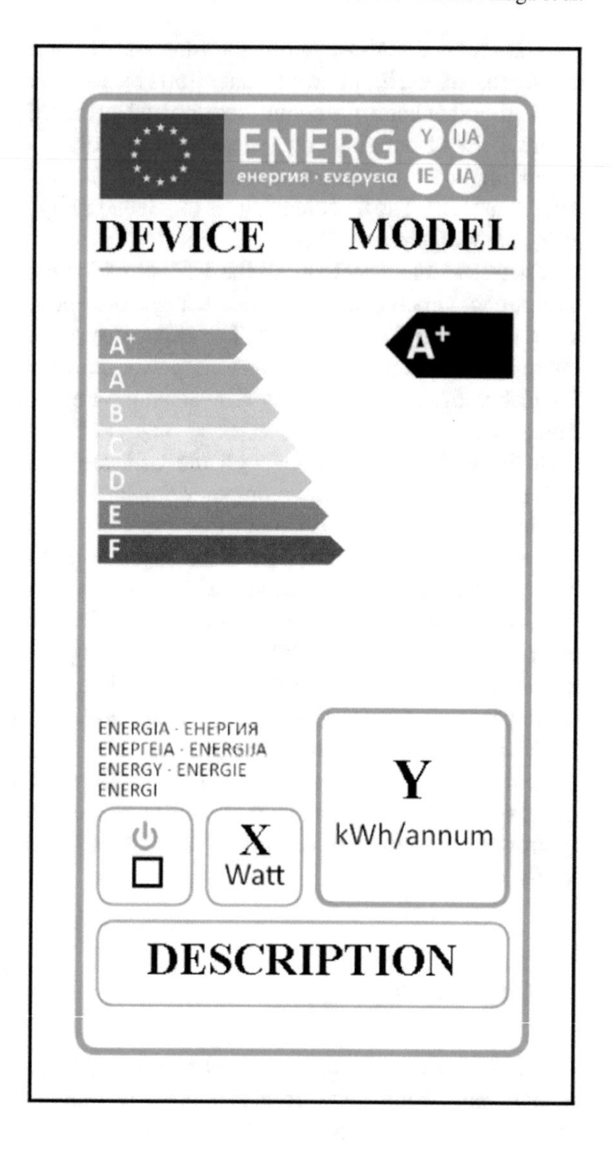


**Fig. 1** Example of classification label on energy efficiency

Such software energy efficiency reference could allow, depending on the users' needs, to choose the most convenient app or library.

The objective of this paper is to present a monitoring tool that allows sample load and energy metrics of software applications and algorithms if there are available sensors to measuring the energy consumed. The energy consumed will be estimated based on the sampled load in other cases. Other future works can use this tool to monitor and classify applications based on their energy consumption. Such classification will help in the challenge of achieving energy-efficient computing.

## 3 Monitoring Tool

The *Phantom monitoring tool* was developed in the European project Phantom [10]. In this paper summarize the important aspects of the Monitoring tool described on the public deliverables [29–32]. It is designed to monitor load, and energy when there available sensors or to estimate it in another case. The metrics sampled will be used for profiling the energy efficiency of the hardware and software, and the monitored status of the computational resources (infrastructure-level), will provide valuable information on the currently available resources. Such information can be used for schedule next execution on the most energy-efficient hardware depending on the user's requirements (energy costs or deadline for complete the execution).

Notice, that the monitoring tool is designed to have not to impact on the system to be monitored. The monitoring tool is composed of a set of monitoring threads, each one in charge of monitoring different kinds of metrics with different sampling frequencies. Basically, the code in the threads is designed to be minimal and keep sleeping most of the time, waking up only for a short period of time for collecting metrics.

The Phantom Monitoring Framework follows the client-server architecture, according to which the runtime monitoring information is collected through a monitoring agent service (deployed on each of the monitored hardware resources) and transmitted to a centralized service (the monitoring server) that is usually deployed on a dedicated resource. Figure 2 shows the client-server architecture of the monitoring framework, which is composed of a server layer where the data is stored, and the monitored environment where the monitoring tool is deployed.

The *Phantom monitoring tool* can Monitor load, and energy when possible, independently on the next two levels:

- **Monitor on system level**: must be able to report the energy consumption independently of each component
- **Monitor on application level**: must be able to monitor each application runs independently, and if possible on algorithm and function levels too

The metrics that can be acquired cover a wide range of functions that target different aspects of the hardware, such as memory, IO, processor, GPU, and network utilization as well as energy consumption. Metrics are composed of a simple key-value and timestamped.

Additionally, the tool estimates the energy consumption when there are not available sensors of energy consumption. The estimation will be done based on the computation load due to the executed application or algorithm and the energy consumption profile of the hardware used in the system.

The Monitor on system level is done by the Monitoring Agent deployed on each device, and the application level is achieved instrumenting the applications with the Monitoring Library. The metrics collected are sent to the Monitoring Server, which provides an API for query and analysis (Fig. 3).

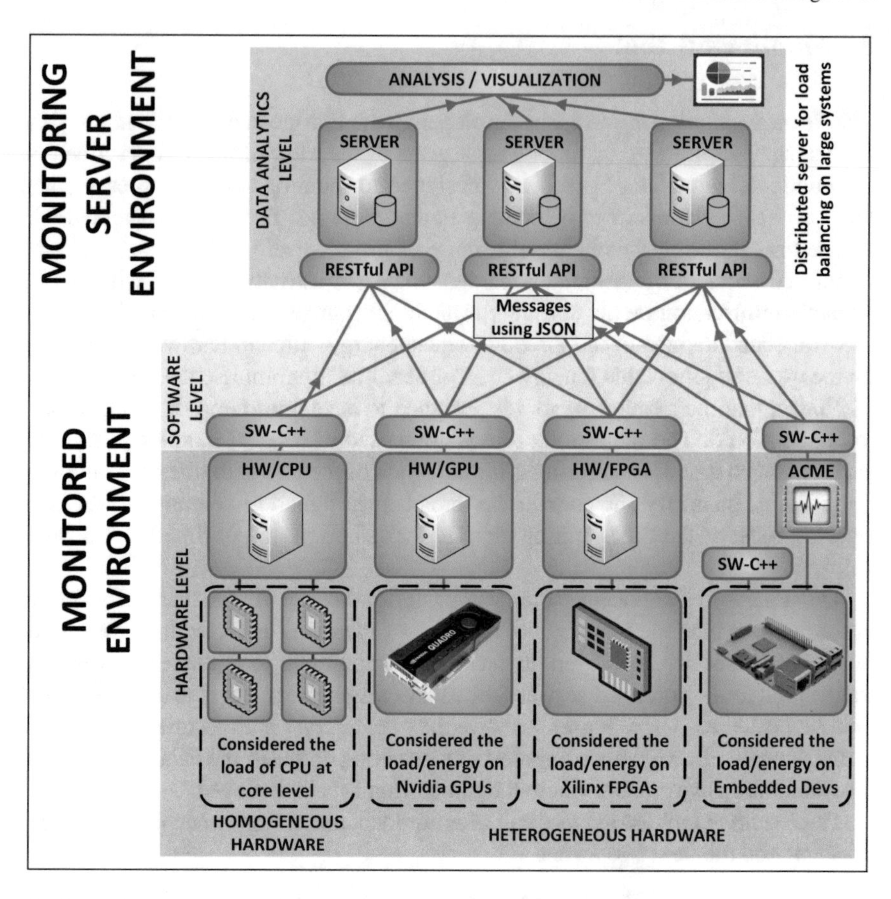


**Fig. 2** Architecture of the monitoring framework

**Monitoring-Agent (monitoring on system level)**:

The design provides a flexible implementation by importing plugins. Only the required plugins are loaded, allowing for the minimization of resources used on the monitored device. Each one of these plugins is an independent thread that wakes up on its sampling time interval, in order to minimize the system utilization. Each thread stores its metrics in an independent local buffer. There is a different time interval for transferring the content of each buffer to the Monitoring Server in other to minimize the network load. The monitoring tool allows that its configuration can be updated during run-time.

The configuration parameters contain a list of metrics to be acquired, their sampling interval, and the frequency at which they should be uploaded to the Server. For simplicity, many default configurations are provided. Figure 4 shows the view of the Web interface.

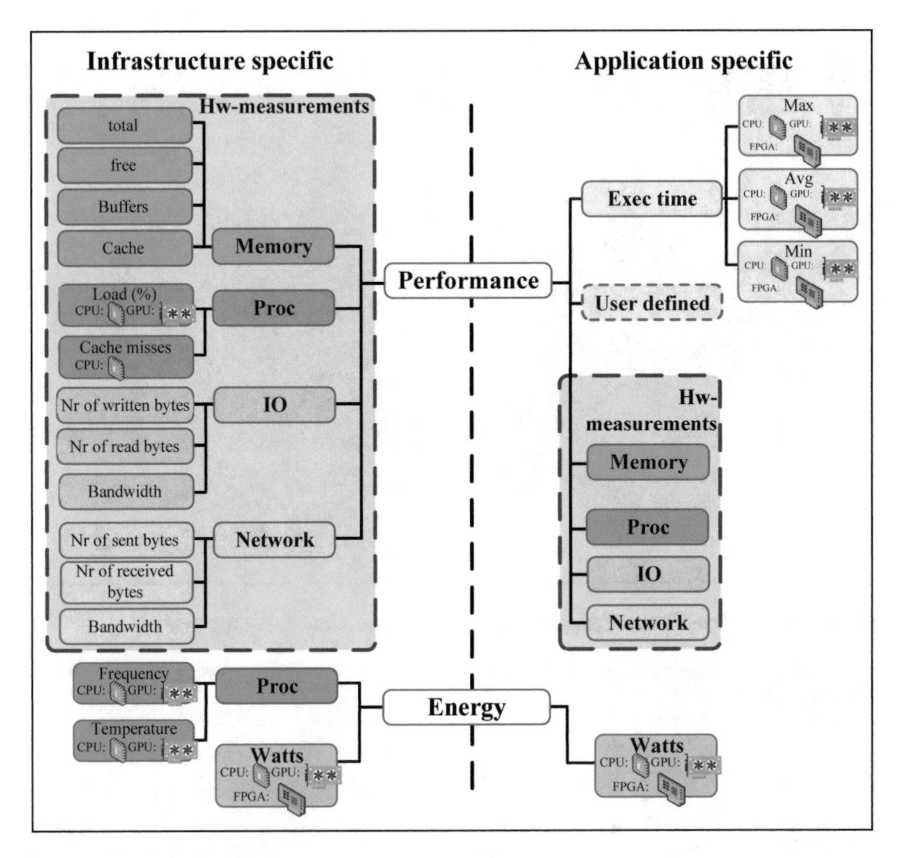


**Fig. 3** Available metrics in the *Phantom monitoring tool*

**Monitoring-Library (monitoring at the application level)**:

The Monitoring Library (application-level monitoring and user-defined metrics collection) abstracts users from the metric collection process. The Phantom Monitoring Library provides a user library and several APIs for instrumentation of the applications. It allows users to control and adjust the metric collection process through user-level APIs. The Monitoring library additionally allows the collection of user-defined metrics. Each user metric is composed of a text label and a value or a set of values, which are automatically timestamped. This allows the users to measure, for example, the execution time of loops or subroutines by registering user-defined measures in the code. Such user-defined metrics are also missing from other existing tools. The code instrumentation allows fine-grained monitoring. Figure 5 shows as an example where the execution time of loops is registered using user-defined metrics in the code.

The instrumented applications with the Monitoring library load when starting their default monitoring configuration. This default configuration, defined by the

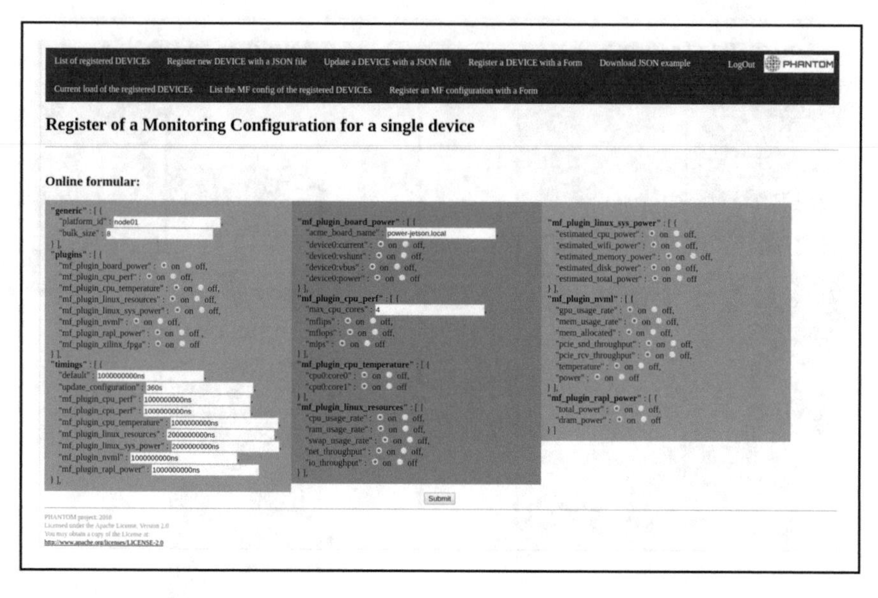


**Fig. 4** Web interface for modifying the monitoring configuration during run-time

```
#include...
extern "C" {#include "../src/mf_api.h"}
...
int main()
{
...
/* MONITORING START */
m_resources.num_metrics = 3;
m_resources.local_data_storage = 1;
m_resources.sampling_interval[0] = 1000; // 1000 ms
strcpy(m_resources.metrics_names[0], "resources_usage");
m_resources.sampling_interval[1] = 1000; // 1000 ms
strcpy(m_resources.metrics_names[1], "disk_io");
m_resources.sampling_interval[2] = 1000; // 1000 ms
strcpy(m_resources.metrics_names[2], "power");
mf_start("192.168.0.8:3033", "node01", &m_resources);
/* simulation process */
for (int n = 0; n < nrLoops; n++) {
auto begin_time = std::chrono::high_resolution_clock::now();
simulation_loop(&(*branches), &(*vertexes), netparams, n, integrationStep);
auto end_time = std::chrono::high_resolution_clock::now();
std::chrono::duration<double, std::milli> duration = end_time-begin_time;
/* MONITORING USER-DEFINED METRICS -> duration of each loop */
sprintf(metric_value, "\%f", duration);
mf_user_metric("duration", metric_value);
}
/* MONITORING END */
mf_end();
/* MONITORING USER-DEFINED METRICS ->total nr. of completed loops */
sprintf(metric_value, "\%d", nrLoops);
mf_user_metric("nrLoops", metric_value);
...
}
```

**Fig. 5** Example of instrumented code with the Monitoring Library

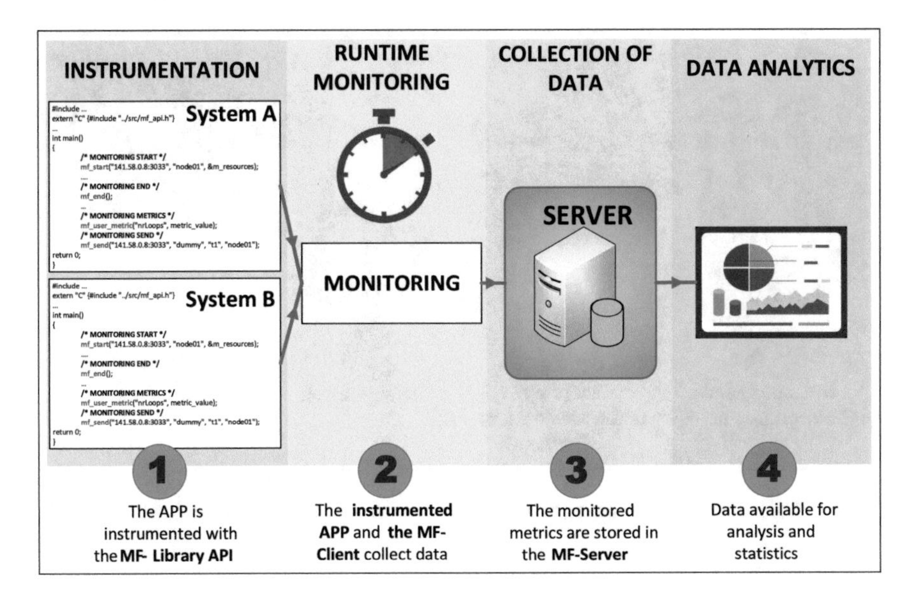


**Fig. 6** Stages of monitoring the execution of an application

system administrator, allows users to not have to perform additional actions and do not need to know the hardware details of the device where the application will run. However, users can, if they wish, provide a different configuration in their instrumented applications.

Figure 6 shows a global view of Monitoring Framework, on the first stage the code is optionally instrumented. The execution of the application and the collection of metrics are done on the second stage. The third stage shows that the collection of metrics can be done during or after the execution of the application. The last stage is when the collected results are analyzed, and the computing efficiency can be calculated.

### Monitoring-Server

The purpose of the server is to store the received metrics and to offer the functionalities to query and analyze the run-time collected metrics. The server is composed of a data storage layer, used to store the monitoring information from the sensors and agents, and a web-service for the data transmission from them to the data storage layer. ElasticSearch was chosen for the data storage component because it is a flexible and powerful real-time search and analytics engine. ElasticSearch was the preferred option because it is a distributed, full-text search engine, and supports RESTful API and web interface, and uses schema-free JSON documents.

The server was implemented with the Node.js runtime environment. Node.js has an event-driven architecture and can be used for data-intensive real-time applications that run across distributed devices.

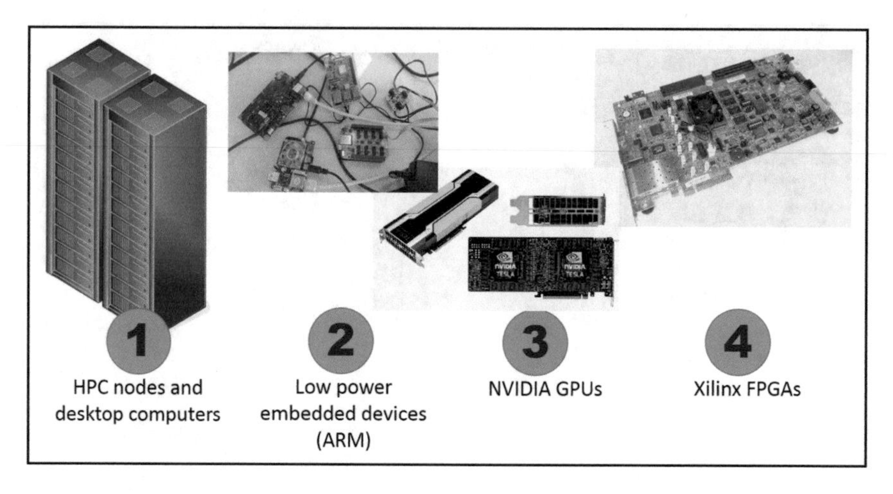


**Fig. 7** Different hardware components monitored with the *Phantom monitoring tool*

## 4 Results

The Monitoring tool was evaluated and validated in three different scenarios:

- **Monitoring embedded devices: Telecom Use Case**
  The use case scenario consists of a microwave radio product from a Telecom industrial partner. In it was monitored a heterogeneous system, composed of a dual-core ARM and ZynQ Xilinx FPGA as processing elements.
- **Monitoring high-performance devices: HPC use case**
  The use case scenario consists of a detailed aerodynamics automotive simulation, provided by HLRS which systems are used for such physical simulations by different industries including the automotive industry. The application implemented with MPI in C was monitored running on HPC computing nodes.
- **Monitoring GPU accelerators: Surveillance use case**
  The use case scenario consists of a Ship Monitoring System solution to maritime situational awareness via processing information from Earth Observation (EO) technologies, such as Satelite images, and data from terrestrial and maritime stations. The parallelized application was monitored running with Intel CPUs and Nvidia GPUs.

Figure 7 shows the different monitored hardware during the evaluation of those scenarios.

The different metrics were accessible in real-time for the partners throw the server web interface and Grafana graphic web interface, as shown in Fig. 8.

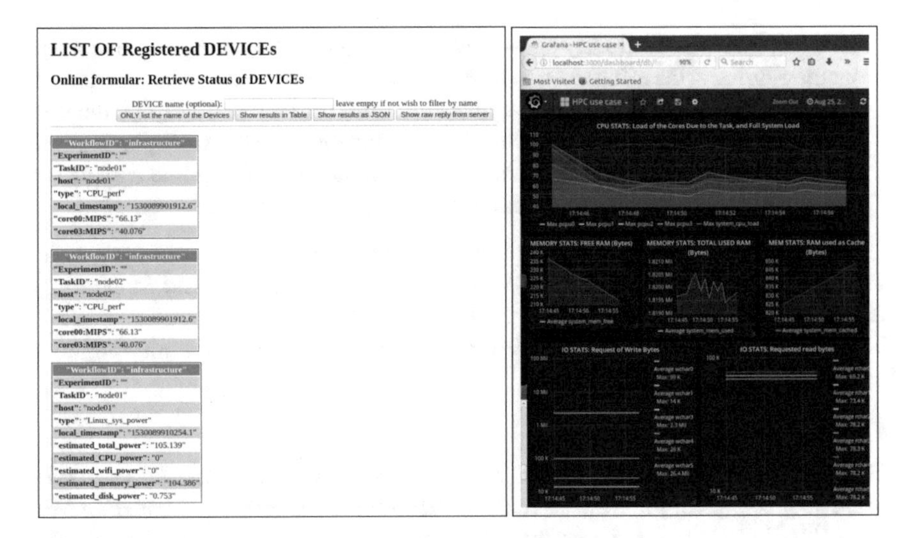


**Fig. 8** Two different available web interfaces for accessing to the registered monitored metrics

# 5 Conclusions

The implemented tool fulfilled all the requirements. It was evaluated with real applications and monitoring overhead on a dual-core AMD processor) showed the additional CPU usage of less than 0.1% and the memory utilization in the order of 5 MB, which is acceptable even in the case of embedded devices.

The developed tool provides major innovations not available on the state-of-the-art monitoring tools in the market:

- **Management of monitoring configurations on heterogeneous systems**. The Phantom Monitoring Framework allows us to register the characteristics of the hardware components of the compute nodes, as well as their default monitoring configuration. This task would normally be done only when nodes are added to the system or when the hardware of the nodes is modified. In this way, the use of the framework is eased to the end-users, since the users don't need to set any configuration, and therefore they don't need to have the expertise to do it.
- **Monitoring of user-defined metrics for the application-level monitoring**. Without the need of installing any additional monitoring tools, application (or user) specific metrics can be monitored. For this purpose, the Monitoring Library offers the users a set of light-weight, hardware-agnostic APIs for the instrumentation of their application code. This allows tight integration of the Monitoring Framework with the user application.
- **Highly customizable monitoring settings**. The Phantom Monitoring Framework was designed with user-friendliness in mind. The users can specify the metrics to be automatically collected, the sampling and transmission periods, and the configuration of each of the sampling metrics, independently for each application, which is not the case for the other, alternative existing solutions.

This configuration flexibility allows defining a more relaxed sampling frequency for monitoring at the infrastructure level (when may not be an application running), and a more high frequency at the application level for those metrics that the user is interested in. The user can define different sample rates for each plugin, even the sampling frequency can be modified during the execution of the application (by modifying the configuration parameters on the web interface).

The source code of Phantom Monitoring Framework is released via GitHub, under Apache V2 license. The source code of the Monitoring-Library and Monitoring-Agent are available online [21]: via the following link:
https://github.com/PHANTOM-Platform/Monitoring/tree/master/Monitoring_client

**Funding**

This work has been done within the projects *Cross-Layer and Multi-Objective Programming Approach for Next Generation Heterogeneous Parallel Computing Systems* (PHANTOM). See the project's web page [10] for further information. The research leading to these results has received funding from the European Unions Horizon 2020 Research and Innovation Programme, grant agreement n. [688146].

**Conflicts of Interest**: The authors declare no conflict of interest. The funders had no role in the design of the study; in the collection, analyses, or interpretation of data; in the writing of the manuscript, or in the decision to publish the results.

## References

1. Top500 list, 2020. [online]. https://www.top500.org/lists/top500/2020/06/. Accessed July 2020
2. Agency, E.P.: EPA ENERGY STAR program requirements for comp. systems. Draft 4 (2009)
3. All4Green Consortium: All4Green. Active collaboration in data centre ecosystem to reduce energy consumption and GHG emissions, Grant agreement ID: 288674
4. Andrae, A.S.G., Edler, T.: On global electricity usage of communication technology: trends to 2030. Challenges **6**(1), 117–157 (2015)
5. Belkhir, L., Elmeligi, A.: Assessing ICT global emissions footprint: trends to 2040 & recommendations. J. Clean. Prod. **177**, 448–463 (2018)
6. Brown, R.E., Masanet, E.R., Nordman, B., Tschudi, W.F., Shehabi, A., Stanley, J., Koomey, J.G., Sartor, D.A., Chan, P.T.: Report to congress on server and data center energy efficiency: public law, pp. 109–431 (2008)
7. CASCADE Consortium: CASCADE. ICT for energy efficient airports, Grant agreement ID: 284920. [online]. https://cascade-eu.org/. Accessed July 2020
8. Consortium, T.E.: Overview of ICT energy consumption (D8.1). Report FP7-2888021. European Network of Excellence in Internet Science (2013)
9. DOLFIN Consortium: DOLFIN. Data Centres Optimization for Energy-Efficient and EnvironmentalLy Friendly INternet, Grant agreement ID: 609140
10. EUXDAT Consortium: PHANTOM. Cross-Layer and Multi-Objective Programming Approach for Next Generation Heterogeneous Parallel Computing Systems, Grant agreement ID: 688146. [online]. https://www.phantom-project.org/. Accessed March 2020

11. Gelenbe, E., Caseau, Y.: The impact of information technology on energy consumption and carbon emissions. Ubiquity **2015**(June), 1–15 (2015)
12. Heddeghem, W.V., Lambert, S., Lanoo, B., Colle, D., Pickavet, M., Demeester, P.: Trends in worldwide ICT electricity consumption from 2007 to 2012. Comput. Commun. **50**(2014), 64–76 (2020)
13. Hintemann, R., Hinterholzer, S.: Energy consumption of data centers worldwide. In: Business, Computer Science (ICT4S) (2019)
14. Horekens, A.: How much $CO_2$ could be worlds most powerful computer save with 2-phase immersion cooling (2013)
15. ICT-Energy Consortium: ICT-energy. Co-ordinating Research Efforts of the ICT-Energy Cmty, G. agreement ID: 611004. [online]. https://www.ict-energy.eu. Accessed July 2020
16. Koomey, J.: Worldwide electricity used in data centers. Environ. Res. Lett. **3**(034008) (2008)
17. Koronen, C., Åhman, M., Nilsson, L.J.: Data centres in future European energy systems-energy efficiency, integration and policy. Energy Effic. **13**(1), 129–44 (2020)
18. Lucas, R., et al.: Top Ten Exascale Research Challenges. Technical Report. U.S. Department of Energy, Office of Science. DEO ASCAC (Advanced Scientific Computing Advisory Committee) Subcommittee Report (2014)
19. Malmodin, J., Moberg, Å., Lundén, D., Finnveden, G., Lovehagen, N.: Greenhouse gas emissions and operational electricity use in the ICT and entertainment & media sectors. J. Ind. Ecol. **14**(5), 770–790 (2010)
20. Masanet, E., Shehabi, A., Lei, N., Smith, S., Koomey, J.: Recalibrating global data center energy-use estimates. Science **367**(6481), 984–986 (2020)
21. Montañana, J.M., Cheptsov, A.: Phantom Monitoring source code. [online]. https://github.com/PHANTOM-Platform/Monitoring/tree/master/Monitoring_client. Accessed July 2020
22. Ou, Z., Pang, B., Deng, Y., Nurminen, J.K., Ylä-Jääski, A., Hui, P.: Energy- and cost-efficiency analysis of ARM-based clusters. In: 12th IEEE/ACM International Symposium on Cluster, Cloud and Grid Computing (ccgrid), pp. 115–123 (2012)
23. Padoin, E.L., Navaux, P.O.A.: High performance computing versus high power consumption. In: IX Workshop de Processamento Paralelo e Distribuído (2011)
24. Pärssinen, M., Kotila, M., Cuevas, R., Phansalkar, A., Manner, J.: Environmental impact assessment of online advertising. Environ. Impact Assess. Rev. **73**, 177–200 (2018)
25. Phillips, A.: ARM Launches Cortex-A50 Series, the World's Most Energy-Efficient 64-Bit Processors (2012)
26. RIKEN Research Institute: Japanese Supercomputer. Fugaku (2020). [online]. https://postk-web.r-ccs.riken.jp/spec.html. Accessed July 2020
27. Rong, H., Zhang, H., Xiao, S., Li, C., Hu, C.: Optimizing energy consumption for data centers. Renew. Sustain. Energy Rev. **58**, 674–691 (2015)
28. The Consortium: DAREED. Decision support Advisor for innovative business models and useR engagement for smart Energy Efficient Districts, G. amt ID: 609082
29. The Consortium: D1.4—Final Design for Cross-layer Programming, Security and Runtime Monitoring (2018). [online]. https://ec.europa.eu/research/participants/documents/downloadPublic?documentIds=080166e5c4cedede&appId=PPGMS. Accessed July 2020
30. The Consortium: D2.2 Final Report on System Software for Multidimensional Optimization on Heterogeneous Systems (2018). [online]. https://ec.europa.eu/research/participants/documents/downloadPublic?documentIds=080166e5bd654e76&appId=PPGMS. Accessed July 2020
31. The Consortium: D3.2 Final Report on Programmer and Productivity-Oriented SW Tools (2018). [online]. https://ec.europa.eu/research/participants/documents/downloadPublic?documentIds=080166e5bd654c3e&appId=PPGMS. Accessed July 2020
32. The Consortium: D4.4 Final Release of Integrated Monitoring Platform, Infrastructure Integration and Resource Management Software Stack (2018). [online]. https://ec.europa.eu/research/participants/documents/downloadPublic?documentIds=080166e5c21dbf9d&appId=PPGMS. Accessed July 2020
33. The Council: The European Economic and Social Committee and the Committee of the Regions: A Digital Single Market Strategy for Europe (2015)

# Speeding Up Vector Engine Offloading with AVEO

**Erich Focht**

**Abstract** Vector Engine Offloading (VEO) was the first implementation of an API for programming the SX-Aurora Tsubasa Vector Engine (VE) like an accelerator, i.e. writing programs for the host CPU which call certain offloading kernels running on the VE. The native VE programming model using OpenMP and MPI still dominates in applications, but CUDA, HIP, OpenMP Target, OpenACC, OpenCL find more and more traction. This report introduces AVEO, an alternative VE offloading implementation with VEO compatible API. It was redesigned to solve a set of problems in VEO and improve call latency as well as memory transfer bandwidth. The results show latency improvements of up to factor 18 and bandwidth increases by factor 8–10 for small buffers and 15–20% for very large buffers. We describe implementation details and remote memory access mechanisms as well as API extensions. This development should contribute to making accelerator-style hybrid programming more attractive on the vector engine, ease porting of hybrid programs but also developing more sophisticated hybrid programming frameworks.

## 1 Introduction

The NEC Aurora Tsubasa Vector Engine (VE) is a very high memory bandwidth vector processor with six HBM2 memory stacks that combines SIMD and pipelining to a power efficient long vector ISA. It was presented at the HotChips 2018 symposium [1] and first performance characteristics were investigated e.g. in [2]. In the variant released in 2020 (VE20) the VE has up to 10 cores with an aggregate memory bandwidth of 1.5TB/s. Each VE is packaged in the form factor of a PCIe card which can be inserted into a x86_64 Linux host computer usually called vector host (VH). Installed setups scale from just one VE in a VH to Infiniband connected clusters of several hundreds of VHes with eight VEs per host.

E. Focht (✉)
NEC Deutschland GmbH, Stuttgart, Germany
e-mail: erich.focht@emea.nec.com



M. M. Resch et al. (eds.), *Sustained Simulation Performance 2019 and 2020*,
https://doi.org/10.1007/978-3-030-68049-7_3

The VE comes with a variety of programming models. Native VE programs written in C, C++ or Fortran run on the VE, parallelized by OpenMP or MPI and offload their system calls to VEOS, the VE operating system executed on the host side. Native VE programs can also offload parts of their code to the host by reverse-offloading [3]. Hybrid MPI programs can run processes on the VE and scalar nodes. And programs running on the host can offload optimized kernels to VEs in a classical accelerator programming model, similar to CUDA [4] or OpenCL [5].

Vector Engine Offloading (VEO) [6] is a framework for enabling users to offload kernels from VH programs to VE. It implements an API that allows to load shared libraries into the VE, locate functions and symbols in them, allocate and free memory on the VE, transfer data to and from the VE as well as asynchronously execute functions on the VE side. It consists of a host side library that interacts closely with the VEOS and a VE binary *veorun helper* that is waiting for commands from the VH and runs on behalf of the user. This framework is necessary because the VE runs without any kernel and is fully controlled from VH side.

VEO has been used to implement heterogeneous programs like the AI optimization framework SOL [7] and Tensorflow-VE [8], provide access to optimized VE libraries from VH-side NumPy Python programs with NLCpy [9], create higher level frameworks for VH-VE heterogeneous programming like HAM [10], implement a CUDA-like device abstraction layer: VEDA [11] or a Python offloading API: PyVEO [12].

The first implementations of VEO from prototype (2015) to version 2.5.0 (2020) were using mechanisms from *libvepseudo*, the core of the VH-side exception and system call handler, to launch and control the veorun helper. The mechanisms used were relying on VE-to-VH system call semantics and tightly coupled the actual host-side user program to a heavily modified version of the *pseudo-process* that normally controls native VE-side programs.

While extending the *pseudo-process* seems the most natural way of implementing VH-VE hybrid programs, the approach brings some hard-to-fix issues:

- The kernel function call latency is large: $O(100\,\mu s)$, which brings a significant handicap to programs with small, short running kernels.
- VH-VE memory transfer speed is limited by the system DMA performance and far below the PCIe x16 generation 3 unidirectional transfer bandwidth.
- Debugging hybrid programs is difficult because one can either attach *gdb* to the host side program or attach the VE *gdb* with some tricks to the VE side program, but never to both sides at the same time.
- The normal performance analysis of the VE side kernels with *ftrace* and *PROGINF* doesn't work out of the box and requires deep system knowledge to enable.
- A VH process can only control one VE, support for multiple VEs would require heavy changes in VEOS.

This work describes the re-implementation of VEO with a completely different approach that solves all above-mentioned issues. The result was named Another/Alternative/Awesome VE Offloading, or short: AVEO.

Section 2 describes the approach taken and the architecture of AVEO, followed by a section on concrete implementation details. Section 4 lists API changes compared to [6]. The paper ends with a section discussing performance results and the conclusion.

## 2 Approach and Architecture

The main motivation for starting this project came from previous work [10, 13] that showed that VH-VE memory transfer bandwidth can be increased by using each core's user DMA descriptors instead of the common system DMA. The difference between the two is that system DMA uses physical memory addresses from unregistered buffers and is initiated from VH side while user DMA requires virtual memory addresses pre-registered on the VE's DMAATB and can only be controlled from VE side. VEO being a VH-side API, the need to control the user DMA engines from VE side complicates the design. In one of the projects, HAM [10], we found that by using a tight loop polling remote memory through user DMA we can actually reduce remote function call latency to a range comparable to CUDA (5–6 μs).

### *2.1 User DMA Communication Mechanisms*

The mechanisms used for transferring data between VH and VE are depicted in Fig. 1. On VH side a system V shared memory buffer allocated in 2MB huge pages is passed to the VE for registration by using *libsysve* VH-VE SHM mechanisms. On VE side a mirror buffer of the same size as the shared memory buffer is allocated and registered.

The VE can read and write directly up to 64 bit words from/to the registered shared memory buffer on the VH-side with the assembler instructions *lhm* and *shm*. This mechanism can be used for polling and controlling mailboxes.

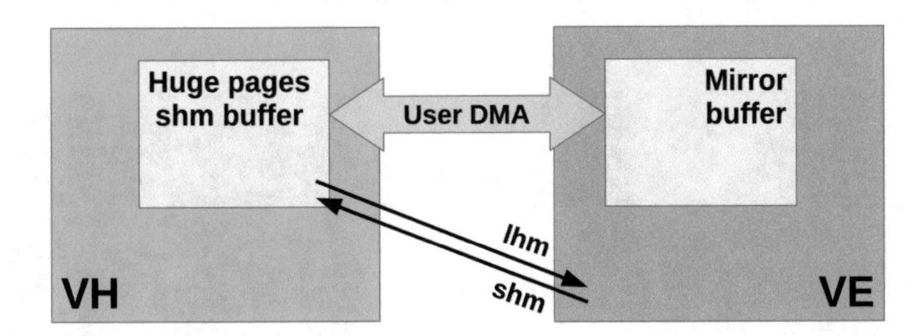


**Fig. 1** VH-VE shared memory mechanisms

With the VE DMA API from *libsysve* the user DMA descriptors can be programmed to transfer memory between the VH shared memory buffer and the mirror buffer (i.e. between two registered buffers) with the help of the VE DMA engines.

The shared memory segment allocated on the VH side can be mapped by any process belonging to the same user. By communicating through it there is no need any more to have the VH-side process play the role of a *pseudo-process* for the VE-side. VH- and VE-side of the VEO code can be decoupled and live in separate processes. This approach solves some of the big issues of the original VEO:

- If VH- and VE-side are different processes then one can attach a debugger to both of them at the same time.
- The VH-side process can communicate with several VE processes, even if they run on different VE cards plugged into the same host. This lifts the restriction of one VE card per VEO process.
- The VE-side process behaves like a normal native VE program and all performance analysis methods just work.
- In principle multiple VE processes could communicate directly with each other through MPI, without involving the VH-side.

## 2.2 VE-URPC

The VE DMA mechanisms are a low level abstraction which is not comfortable to use in complex programs, therefore we decided to implement a Remote Procedure Call (RPC) component using the VE user DMA. It was named *VE-URPC* and its architecture is sketched in Fig. 2. It consists logically of two queues or pipes that are connecting the VH and the VE, one for sending and one for receiving commands. The receiving end of a pipe is attached to a list of handlers where each command has registered the appropriate processing functions.

A command (yellow) and its payload is pushed into the send queue on the VH (1), transferred to the VE where it is processed by the proper handler function attached to the receiving end of the queue (2). The result is also a command (green) that

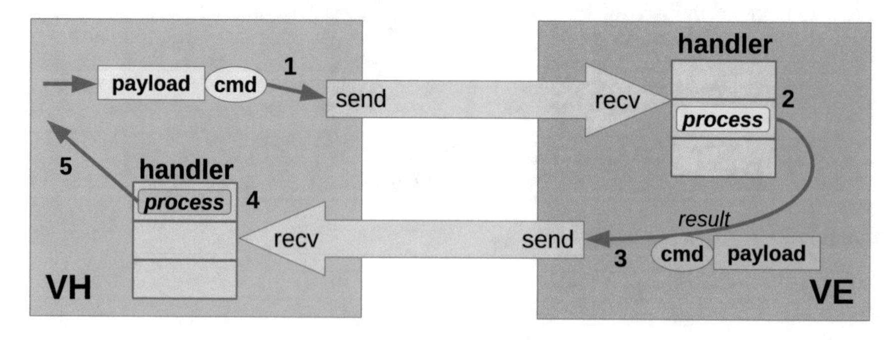


**Fig. 2** VE-URPC architecture and path of a command through the system

gets inserted into the VE's send queue and is processed on VH side by the handler corresponding to the reply command (4). Finally the handler's result is delivered to the command issuer as the result of the RPC.

A VE-URPC instance connects exactly two peers and requires an allocated buffer, either huge pages system V shared memory on VH side, or a mirror buffer on VE side. A VH connected to multiple VEs requires several instances of VE-URPC and therefore several buffers.

## 2.3 AVEO on VE-URPC

With VE-URPC we can build a VE Offloading framework that is fully compatible to the original described in [6]. The architecture shown in Fig. 3 is very similar to VEO's but has important differences in details related to processes and threads:

- A *ProcHandle* instance allocates the VH side of a VE-URPC buffers, starts an *aveorun* helper VE process which connects to the *ProcHandle*'s VE-URPC instance.
- The first VE-URPC instance of a *ProcHandle* is linked to the first/default *Context*.
- Unlike original VEO, a *Context* does not do syscall handling for the VE side and does not require an additional thread on VH-side.
- Additional *Contexts* require additional VE-URPC instances and connect to clones of the main aveorun thread on VE-side. They don't require additional threads on VH-side.
- The syscall-handlers of the *aveorun* helper threads on VE-side are the normal pseudo-processes of native VE processes, thus need no adaptation to hybrid programming.
- Any number of additional *ProcHandle* instances can be created within a VH program, targeting any of the VEs in the system.

In principle the VE processes could communicate directly with each other but this feature is not implemented in the current AVEO design.

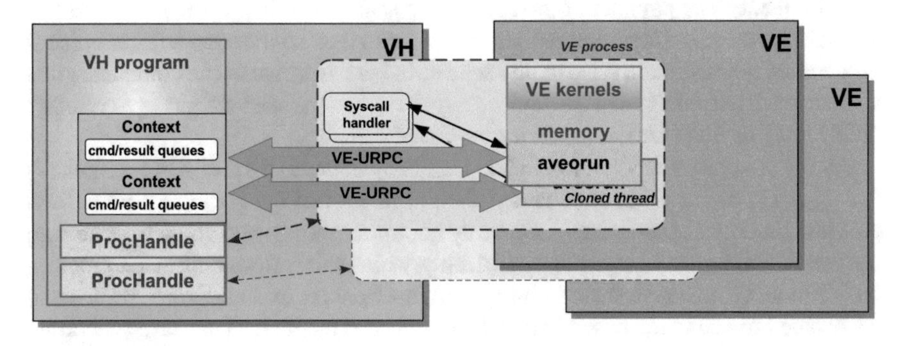


**Fig. 3** AVEO architecture

# 3 Implementation

The memory layout of one of the two VE-URPC transfer queues living on the VH side in the shared memory buffer is described in Listing 1.

```
1 struct transfer_queue {
2         volatile uint32_t sender_flags;
3         volatile uint32_t receiver_flags;
4         volatile int64_t last_put_req;
5         volatile int64_t last_get_req;
6         volatile urpc_mb_t mb[URPC_LEN_MB];
7         volatile uint64_t data[DATA_BUFF_END / sizeof(uint64_t)];
8 };
```

**Listing 1** Structure of one VE-URPC transfer buffer.

The *sender_flags* and *receiver_flags* are mainly reserved for future use, currently just used for notifying the other end whether we're ready with initialization. In the *last_put_req* the sender marks which VE-URPC request ID he last wrote. The request's slot is calculated from the ID and this field is only modified by the sender. The *last_get_req* is only modified by the receiver who marks in it which request ID was the last one read. The *mb* array contains 256 command slots which are filled and processed in round-robin order.

A VE-URPC command slot has 64 bits, 8 bits are used for the command ID which can take values between 1 and 255. The command ID is set to zero by the receiver once the command has been executed and its payload is not needed any more. A payload for the command is stored in the *data* array, 29 bits in the command are used to pass the offset of the payload in this buffer. And the last 27 bits of the command encode the length of the payload, which can be up to 128 MiB and should be transferrable in a single DMA transaction.

VE-URPC requests are identified by an ID which is incremented for every request. Every command has to send a reply and the IDs of the replies are kept in sync with the IDs of their commands. Up to 256 requests can be "in-flight" on a VE-URPC transfer queue. The payload of the commands is allocated inside the *data* buffer of the transfer queue using a simple algorithm that takes contiguous adjacent pieces of memory. Allocation is done by the sender, the receiver marks the commands that were finished and don't need the payload any more, and the sender "garbage-collects" the freeable payload buffers when it runs out of memory.

AVEO is implemented on top of VE-URPC by defining a set of RPC commands and replies (which are also commands), their handlers, and integrating the VE-URPC progress functions appropriately. One way of implementing aveo could have been to use two RPCs, a generic async call and a reply, which call various builtin functions on the VE side, eg. to load a shared library, find functions, or execute functions. Instead we decided to use many specialized RPCs with own handlers. That saves decoding time because it needs less arguments and simplifies debugging. Table 1 shows a list of the RPCs implementing AVEO.

**Table 1** VE-URPC commands implementing AVEO

| RPC ID | Command | Function |
|---|---|---|
| 0 | URPC_CMD_NONE | No command, marks that slot is done |
| 1 | URPC_CMD_PING | Alive check, replies with ACK |
| 2 | URPC_CMD_EXIT | Signal peer to exit |
| 3 | URPC_CMD_ACK | Result with no content, void return type |
| 4 | URPC_CMD_RESULT | Simple result (int64_t) without stack frame |
| 5 | URPC_CMD_RES_STK | Result with stack frame as payload |
| 6 | URPC_CMD_EXCEPTION | Exception notification |
| 7 | URPC_CMD_LOADLIB | Load .so shared object |
| 8 | URPC_CMD_UNLOADLIB | Unload .so shared object |
| 9 | URPC_CMD_GETSYM | Find symbol in loaded shared object |
| 10 | URPC_CMD_ALLOC | Allocate buffer on VE |
| 11 | URPC_CMD_FREE | Free buffer on VE |
| 12 | URPC_CMD_SENDBUFF | Request sending a buffer as payload, args contain address and length |
| 13 | URPC_CMD_RECVBUFF | Request receiving a buffer in the payload, args contain destination address |
| 14 | URPC_CMD_CALL | Call a VE function with no stack transfer |
| 15 | URPC_CMD_CALL_STKIN | Call a VE function with stack "IN" transfer |
| 16 | URPC_CMD_CALL_STKOUT | Call a VE function with stack "OUT" transfer |
| 17 | URPC_CMD_CALL_STKINOUT | Call a VE function with stack "IN" and "OUT" transfer |
| 18 | URPC_CMD_SLEEPING | Reserved for future use |
| 19 | URPC_CMD_NEWPEER | Create new remote peer (context) inside same process |

VE-URPC communication progress is implemented with following constraints in mind: User DMA is driven from the VE side, and context switches are expensive on VE side. Sending a command is equivalent to putting it into the *mb* mailbox array and its payload into the *data* array of the "send" transfer queue. The actual message transfers and execution of command handlers happen in VH and VE specific receive progress functions. On VH-side this progress function is invoked after issuing commands and when waiting for them to finish. On VE side the progress function also triggers DMA programming and proper data transfer, and is called in a tight loop.

The *aveorun* helper running on VE side reports expectations to the VH by sending the *URPC_CMD_EXCEPTION* from an exception handler. The exceptions that are reported are SIGABRT, SIGFPE, SGILL, SIGSEGV. All of them are fatal for the VE side and cannot be recovered, therefore they lead to termination of *aveorun*.

## 4 API Extensions

While AVEO aims to be fully compatible with the original VEO API [6], the change also brings the opportunity to extend the API and test new features. This section describes API extensions available in AVEO version 0.9.11.

### Specify the core of a ProcHandle instance

The VE core on which a proc shall run can be specified by setting the environment variable

```
export VE_CORE_NUMBER=<core>
```

The ProcHandle instance will start the *aveorun* helper for its first/default context on the specified core.

### Find the contexts of a ProcHandle

A *ProcHandle* instance can have multiple contexts and their number can be retrieved by the function

```
int veo_num_contexts(struct veo_proc_handle *proc);
```

Pointers to context structures can be retrieved by calling:

```
struct veo_thr_ctxt*
veo_get_context(struct veo_proc_handle *proc, int idx);
```

with *idx* taking values between 0 and the result of *veo_num_contexts(proc) - 1*. The context with $idx = 0$ is the main context of the *ProcHandle*. It will not be destroyed when closed, instead it is destroyed when the *proc* is killed by calling *veo_proc_destroy()*, or when the program ends.

### Unloading a VE library

VEO loads VE libraries but can not unload them. In AVEO the call

```
int veo_unload_library(struct veo_proc_handle *proc,
                       const uint64_t libhandle);
```

unloads a VE library and removes its cached resolved symbols. This saves memory but also allows to implement simple JIT compilation on top of AVEO.

### Context synchronization

The function

```
void veo_context_sync(veo_thr_ctxt *ctx);
```

will return when all requests in the passed context's command queue have finished processing. Note that their results still need to be picked up by either *veo_call_peek_result()* or *veo_call_wait_result()*.

**Synchronous kernel call**

A synchronous kernel call is a shortcut of an asynchronous call and waiting for its result. In AVEO it is implemented differently from an asynchronous call and is being executed on a *proc* instead of a context. It can be used without actually having opened any context.

```
int veo_call_sync(struct veo_proc_handle *h, uint64_t addr,
                  struct veo_args *ca, uint64_t *result);
```

## 5 Performance and Usage

The core motivation for AVEO was to reduce the call latencies that were limiting VEO's usability. In order to verify the results of the development we did measurements on a SX-Aurora Tsubasa A300-2 system with one Skylake Gold 6126 CPU and two VE10B vector engines running at 1.4 GHz.

The call latency results are shown in Figs. 4 and 5: the old, VEO latencies in the left plot and the new, AVEO latencies in the right plot, with latencies being computed from the time it took to do a certain number of calls and wait for their result. The time measurements were done in four ways:

1. Test 1: submit N asynchronous calls then wait for each of the N requests to finish by issuing *veo_call_wait_result()*.
2. Test 2: submit one asynchronous call then wait for its result, repeat N times.
3. Test 3: submit N asynchronous calls then wait only for the last request to finish by issuing *veo_call_wait_result()*. Ignore the other requests' results.
4. Test 4: submit N synchronous calls. This test can not be run on VEO.

The VEO results show the expected latency in the order of magnitude of 100 $\mu$s, closely tied to the system call latency. The green curve of test 3 unexpectedly starts to increase latency with growing number of calls. It turned out that this effect results from an implementation detail, a linear search in the results queue that uses significant time when the number of requests is large. In AVEO the implementation was fixed to avoid the linear search.

AVEO call latencies go down to 5.4 $\mu$s with test 3 showing the best results, closely followed by test 1. Tests 2 and 4 are slightly above 6 $\mu$s.

Memory transfer latencies are influenced by the call latencies, too. By transfering 30000 times one byte we measured the latency of one transfer for both VEO and AVEO. The result is shown in Table 2 and represents an improvement of factor 16 for the VH to VE transfer and 8 for VE to VH.

The memory transfer of payloads and buffers between host and device is done through User DMA through buffers that are registered with the VE's DMAATB. That means: such a memory transfer involves chopping the buffer into appropriate pieces that fit the transfer buffer, copying it from its original location into the VH or

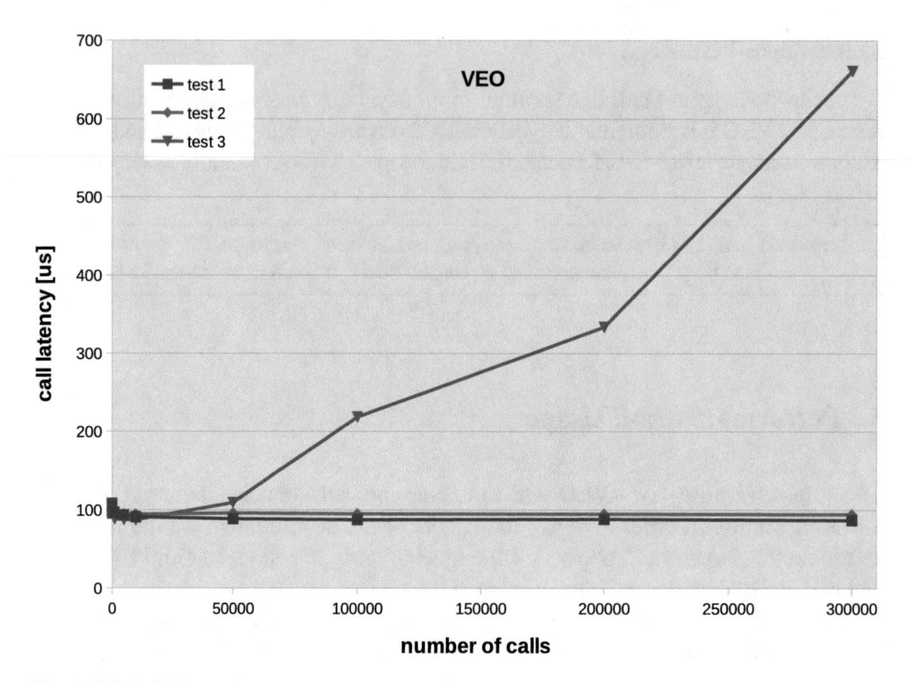


Fig. 4 VEO call latencies

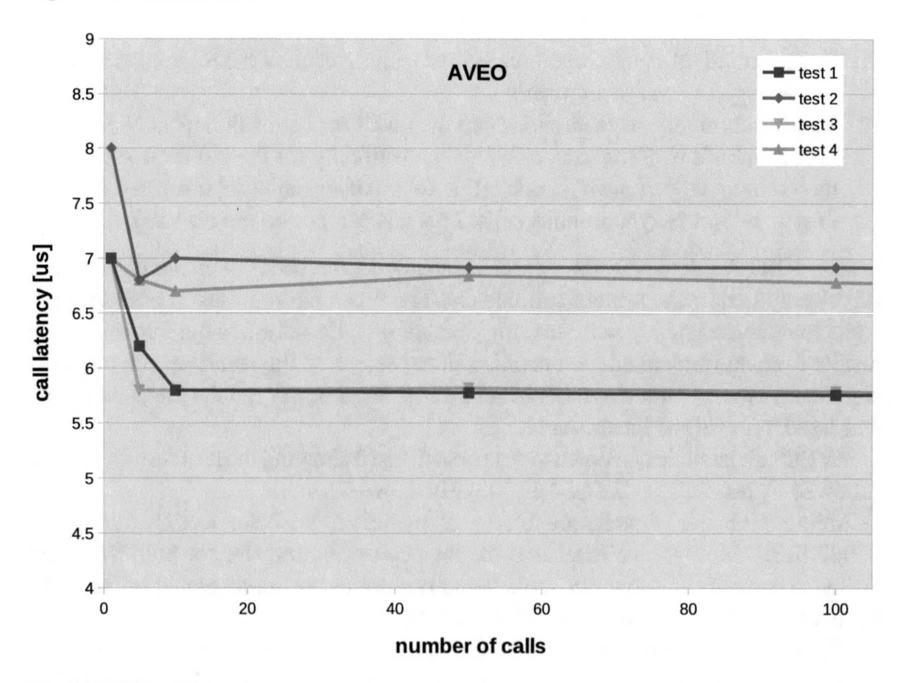


Fig. 5 AVEO call latencies

**Table 2** Latency of 1 byte transfers from host to device memory (veo_write_mem) and device to host memory (veo_read_mem)

| Call | VEO (μs) | AVEO (μs) |
|---|---|---|
| veo_write_mem | 97.3 | 6.03 |
| veo_read_mem | 56.1 | 7.09 |

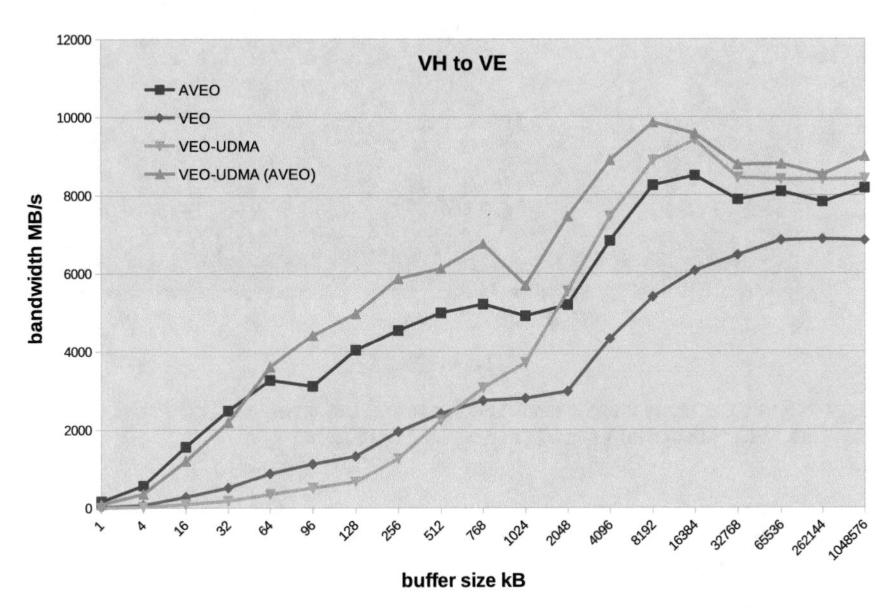


**Fig. 6** VH to VE transfer bandwidths for buffer sizes between 1 kB and 1 GB, measured for AVEO, VEO, VEO-UDMA and VEO-UDMA on top of AVEO. The x-axis is listing the measured buffer sizes in incremental order, and therefore only piece-wise logarithmic

VE transfer buffer, moving it between VH and VE with the help of the DMA engine and copying the buffer to the destination location. Of course one could imagine a more direct way of transfer if the VH-side program would allocate its data in buffers which are mappable on the VE side, which currently need to be shared memory pages in 2MB huge pages. The VE-side would itself need to allocate its data on mapped memory. While possible, it involves own allocators and large amounts of huge pages reserved on the VH side. This was left for a later exercise.

The current AVEO implementation's transfer bandwidths were compared to those measured with VEO and with the add-on VEO-UDMA [13] built on top of VEO and AVEO. VEO-UDMA uses a separate set of buffers for user DMA and overlaps DMA and memcopy operations.

The results are plotted in Figs. 6 and 7. On the low buffer size end AVEO (blue curve) is outperforming VEO by factors of 8–10. At the largest buffer sizes it still maintains an advantage of 15–20%. VEO-UDMA on top of AVEO benefits from the latency gain at small buffer sizes but its overhead is larger than AVEO's, therefore

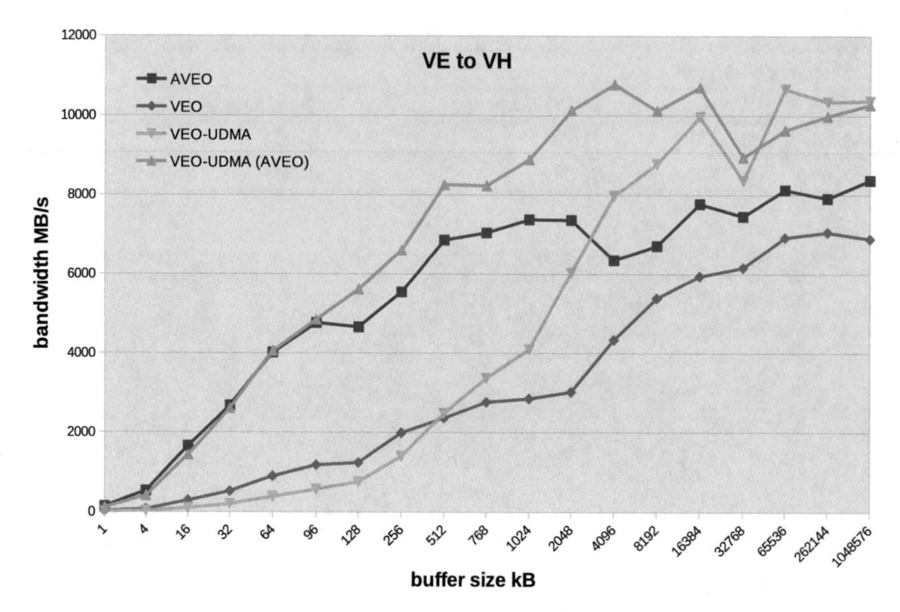


**Fig. 7** VE to VH memory transfer bandwidths for buffer sizes between 1 kB and 1 GB, measured for AVEO, VEO, VEO-UDMA and VEO-UDMA on top of AVEO

it has slightly lower performance up to 64–96 kB buffers. It wins clearly for buffer sizes larger than 96 kB.

## 6 Conclusion

The SX-Aurora Tsubasa Vector Engine primary usage and programming mode was that of native VE-only programs. With VEO [6] we implemented an API that allowed to write hybrid programs treating the VE as an accelerator. Since then VEO was picked up by several software projects and served as foundation to some hybrid programming infrastructures for the VE, for example in the development of an OpenMP Target Device Offloading [14], but also showed limitations in performance and capabilities which were hard to overcome within its core design.

This paper has introduced AVEO, an alternative to VEO with fully compatible API. With changed architecture it solves all notable problems of VEO like debugging, performance profiling, multiple VE usage, and improves significantly the call latency and memory transfer bandwidths. The performance improvements were shown in experiments comparing with VEO. The currently used method of remotely polling a mailbox located in shared memory on host side leaves little room for the improvement of the latency, which is currently in the range of 5.5–6 μs. The memory transfer bandwidth can still be increased by communicating directly between DMAATB registered buffers and eliminating the *memcpy* operations on host and device.

AVEO's API extensions were also briefly introduced, the simplified architecture allows for easier development of such changes. We are aiming at improving synchronization between host and device, possibly adding mechanisms for direct device to device transfers and integrating with MPI.

## References


1. Yamada, Y., Momose, S.: Vector engine processor of NEC's brand-new supercomputer SX-Aurora TSUBASA. In: Hot Chips Symposium on High Performance Chips (August 2018), https://www.hotchips.org [last visited: 05/19]
2. Komatsu, K., Momose, S., Isobe, Y., Watanabe, O., Musa, A., Yokokawa, M., Aoyama, T., Sato, M., Kobayashi, H.: Performance evaluation of a vector supercomputer SX-Aurora TSUBASA. In: Proceedings of the International Conference for High Performance Computing, Networking, Storage, and Analysis (SC '18), Article 54, 12 pp. IEEE Press, Piscataway, NJ, USA. https://doi.org/10.1109/SC.2018.00057
3. Libsysve documentation: getting started with VHcall. https://veos-sxarr-nec.github.io/libsysve/md_doc_VHCall.html
4. Nickolls, J., Buck, I., Garland, M., Skadron, K.: Scalable parallel programming with CUDA. ACM Queue **6**(2), 40–53 (2008)
5. Stone, J.E., Gohara, D., Shi, G.: OpenCL: a parallel programming standard for heterogeneous computing systems. IEEE Comput. Sci. Eng. **12**(3), 66–73 (2010)
6. Focht, E.: VEO and PyVEO: vector engine offloading for the NEC SX-Aurora Tsubasa. In: Resch, M., Kovalenko, Y., Bez, W., Focht, E., Kobayashi, H. (eds.) Sustained Simulation Performance 2018 and 2019, pp. 95–109. Springer International Publishing (2020)
7. Weber, N.: Sol: transparent neural network acceleration platform. In: Proceedings of Supercomputing 2018. https://sc18.supercomputing.org/proceedings/tech_poster/poster_files/post142s2-file3.pdf
8. TENSORFLOW-VE github repository. https://github.com/sx-aurora-dev/tensorflow
9. NLCpy project github repository. https://github.com/sx-aurora/nlcpy
10. Noack, M., Focht, E., Steinke, T.: Heterogeneous active messages for offloading on the NEC SX-Aurora TSUBASA. Proc. IPDPSW 26–35 (2019). https://doi.org/10.1109/IPDPSW.2019.00014
11. VEDA: vector engine device API github repository. https://github.com/SX-Aurora/veda
12. PyVEO github repository. https://github.com/SX-Aurora/py-veo
13. VEO-UDMA project github repository. https://github.com/sx-aurora/veo-udma
14. Cramer, T., Römmer, M., Kosmynin, B., Focht, E., Müller, M.: OpenMP target device offloading for the SX-Aurora TSUBASA vector engine. In: Wyrzykowski, R., Deelman, E., Dongarra, J., Karczewski, K. (eds.) Parallel Processing and Applied Mathematics, PPAM 2019. Springer International Publishing (2020)

# Numerics and Optimization

# Optimizations of DNS Codes for Turbulence on SX-Aurora TSUBASA

**Yujiro Takenaka, Mitsuo Yokokawa, Takashi Ishihara, Kazuhiko Komatsu, and Hiroaki Kobayashi**

**Abstract** Direct numerical simulations (DNSs) of incompressible turbulence have been performed since the late 1960s, but simulations that reproduce strongly nonlinear turbulent flows as in the real-world have not been realized. We have implemented two kinds of parallel Fourier-spectral DNS codes by using a one-dimensional domain decomposition (slab decomposition) and a two-dimensional domain decomposition (pencil decomposition) for a cutting-edge vector supercomputer in order to carry out larger DNSs than ever before. In the DNS by the Fourier spectral method, the three-dimensional Fast Fourier Transforms (3D-FFTs) account for more than 90% of the computational time. Thus, in this article, our FFT codes for vector computers are optimized on SX-Aurora TSUBASA, and vector execution performance of the codes is measured. After optimization, the calculation time of the pencil decomposition code is 1.6 times shorter than before optimization.

Y. Takenaka (✉) · M. Yokokawa
Graduate School of System Informatics, Kobe University, Kobe, Japan
e-mail: ytakenaka@stu.kobe-u.ac.jp

M. Yokokawa
e-mail: yokokawa@port.kobe-u.ac.jp

T. Ishihara
Graduate School of Environmental and Life Science, Okayama University, Okayama, Japan
e-mail: takashi_ishihara@okayama-u.ac.jp

K. Komatsu
Cyberscience Center, Tohoku University, Sendai, Japan
e-mail: komatsu@tohoku.ac.jp

H. Kobayashi
Graduate School of Information Sciences, Tohoku University, Sendai, Japan
e-mail: koba@tohoku.ac.jp



M. M. Resch et al. (eds.), *Sustained Simulation Performance 2019 and 2020*,
https://doi.org/10.1007/978-3-030-68049-7_4

# 1 Introduction

Turbulence exists everywhere around us, and it is important to elucidate its property. Although direct numerical simulations (DNSs) of incompressible turbulence have been performed since the late 1960s [1–4], simulations that reproduce strongly nonlinear turbulent flows as in the real-world have not been realized. The DNS with $12{,}288^3$ grid points on the K computer was performed, in which the Taylor scale Reynolds number is approximately 2,300 [3, 5]. The DNS using GPUs with $18{,}432^3$ grid points on the Summit at Oak Ridge National Laboratory was reported [4].

We developed parallel Fourier-spectral DNS codes by using a one-dimensional domain decomposition (slab decomposition, referred to as "slab code") and a two-dimensional domain decomposition (pencil decomposition, referred to as "pencil code") by using our Fast Fourier Transform (FFT) library. In the codes, most of the computational time is consumed in calculating the three-dimensional FFTs (3D-FFTs) that cause data transfer among processors. The method of data transfer among processors is different between the slab code and the pencil code. In addition, many of the same operations to 3-dimensional large array variables for the Fourier coefficients, or vector operations, are preformed in the FFT routines.

The vector computers have vector processors that can perform the same arithmetic operations for vector data at high speed. Thus, it is suitable for scientific computation such as fluid simulations.

This article describes the optimizations for the slab code and the pencil code on the vector supercomputer SX-Aurora TSUBASA. In the slab code, we could reduce the CPU port conflict time by exchanging the order of the loop. In the pencil code, we optimized the cache hit ratio by loop blocking for the transposition calculation in one core.

# 2 Parallel Direct Numerical Simulation

## 2.1 *Basic Equations and Spectral Method*

The flow of incompressible fluid as described by the Navier–Stokes equations is considered,

$$\frac{\partial \boldsymbol{u}}{\partial t} + (\boldsymbol{u} \cdot \nabla)\boldsymbol{u} = -\frac{1}{\rho}\frac{\partial p}{\partial \boldsymbol{x}} + \nu \Delta \boldsymbol{u} + \boldsymbol{F}, \tag{1}$$

under a periodic boundary condition with period $2\pi$, where $\boldsymbol{u} = (u, v, w)$ is the velocity field, $p$ is the pressure, and $\boldsymbol{F}$ is the external force. The fluid density is assumed to be unity. The puressure term on the right hand side of Equation (1) can be eliminated by the incompressibility condition:

$$\nabla \cdot \boldsymbol{u} = 0. \tag{2}$$

Applying the Fourier spectral method to Eqs. (1) and (2), we obtain the following equation:

$$\frac{\mathrm{d}}{\mathrm{d}t}\hat{\boldsymbol{u}}_k = -\nu|\boldsymbol{k}|^2\hat{\boldsymbol{u}}_k + \boldsymbol{k}\cdot\frac{\boldsymbol{k}\cdot(\widehat{(\boldsymbol{u}\cdot\nabla)\boldsymbol{u}})_k}{|\boldsymbol{k}|^2} - (\widehat{(\boldsymbol{u}\cdot\nabla)\boldsymbol{u}})_k, \tag{3}$$

where $\boldsymbol{k}$ is the wave vector, and the hat $\hat{}$ denotes the Fourier coefficient. Here, $\hat{\boldsymbol{u}}_k = (\hat{u}, \hat{v}, \hat{w})_k$ is defined by

$$\boldsymbol{u}_j = \sum_{k<k_{max}} \hat{\boldsymbol{u}}_k \mathrm{e}^{-\mathrm{i}k\cdot\frac{j}{2\pi}}.$$

The convolutional summation is needed to perform an evaluation of the second term on the right hand side of Eq. (1) that is a nonlinear term. However, by using FFT, the product could be calculated in the real space, that is, first, the Fourier inverse transforms are applied to $\hat{u}$, $\hat{v}$, and $\hat{w}$ to obtain the velocities in the real space. Then, the products $uu$, $uv$, $uw$, $vv$, $vw$, and $ww$ are transformed to spectral space by the Fourier transforms. However, aliasing errors appear in the convolution summation calculations. Therefore, the errors are removed by using a phase shift method and cutting the Fourier modes of $k > k_{max}$. In a phase shift method, the nonlinear term needs to be evaluated twice, thus 18 3D-FFTs are needed in total for a calculation of the nonlinear term.

The 4-th order Runge–Kutta–Gill method is used for advancing time.

## 2.2 *Parallelization*

In the DNS code by the Fourier spectral method, the 3D-FFT calculation part accounts for more than 90% of the computational time. Our original 2 and 4-radix FFT routines are used instead of using any open FFT libraries so that parts of calculations before and after 3D-FFT calls should be merged into the first and last part of 3D-FF calculations. In a parallel implementation, the 3D-FFT has a global data dependence because of its requirement for global summation over all processors. Thus, data transfer is needed during the computation.

In the codes, domain decompositions are used. In the slab code, one global transposition among processors is required in one 3D-FFT to have data continuously in one processor after 1D-FFT is performed in each of the two dimensions (Fig. 1). Let $N^3$ be grid points and $n_p$ be the number of processors, $(N \times N \times (N/n_p))/n_p$ data are transferred one time each 3D-FFT among $n_p$ processors.

In the portion of the 3D-FFT, the first element of the three-dimensional array is accessed continuously to increase the vector length. The CPU port conflict time becomes large because its access sequences to the memory become non-contiguous. Since the port conflict time cannot be ignored to get higher performance, it needs to be reduced by exchanging the order of the loop.

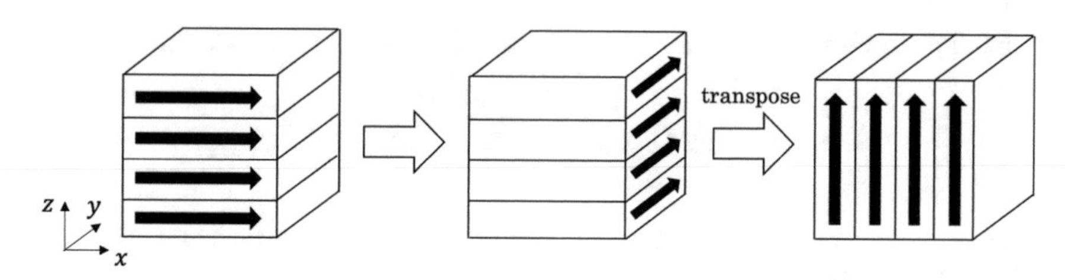


**Fig. 1** One transposition is required in 3D-FFT procedure by the slab decomposition

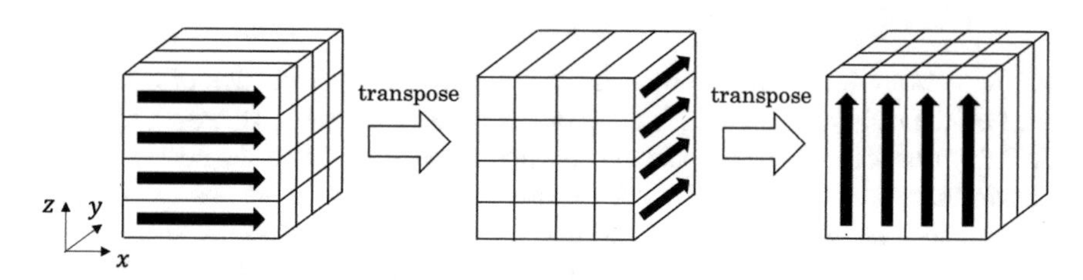


**Fig. 2** Two transpositions are required in 3D-FFT procedure by the pencil decomposition

In the pencil code, one processor has data continuously only in one dimension. Thus, it is necessary to perform global transposition after every 1D-FFT is performed (Fig. 2). After performing the x-direction Fourier transform, a transposition is performed, and after the y-direction Fourier transform is performed, a transposition is performed, and finally the z-direction Fourier transform is performed. Two transpositions in one 3D-FFT are required. Since each transposition can be done independently, we split a whole computational domain into several slabs by `MPI_Comm_split` to confine the message in the slab, creating multiple communicators. $(N \times (N/\sqrt{n_p}) \times (N/\sqrt{n_p}))/\sqrt{n_p}$ data are transferred two times among $\sqrt{n_p}$ processors in each split slab for each 3D-FFT.

The computational time of the portion of the transposition is longer than that of the slab code because the average vector length is small. The computational time could be reduced by increasing the cache hit ratio instead of increasing the vector length. In order to increase the cache hit ratio, it is effective to split the loops by loop blocking.

# 3 Performance Evaluation

## *3.1 SX-Aurora TSUBASA*

Performance of the DNS codes is measured on the vector computer SX-Aurora TSUBASA [6]. The SX-Aurora TSUBASA is a parallel computer system of the distributed memory type and consists of a vector host (VH) and one or more vector engines (VEs). The VH is a Linux server that provides standard operating system

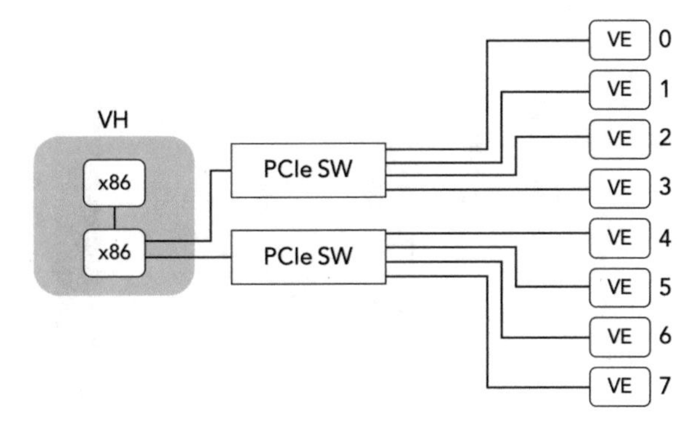


**Fig. 3** An A300-8 system has 8 VEs via 2 PCIe connections to the VH

**Table 1** The specification of a VE of SX-Aurora TSUBASA

| | Type 10B |
|---|---|
| Number of cores | 8 |
| Peak performance | 2.15 TFlops |
| Memory bandwidth | 1.22 TB/s |
| L1 data cache | 32 KB |
| L1 instruction cache | 32 KB |
| L2 cache | 256 KB |
| L3 cache | 16 MB |

(OS) functions. The VE is built as a Peripheral Component Interconnect Express (PCIe) card, on which a vector processor is mounted. VEOS is an OS for VEs that runs on the VH and controls VEs. We used the A300-8 Type 10B system, thus there are 8 VEs via two PCIe connections (Fig. 3). Each VE has a 16 MB last-level cache (LLC) and its performance is 268.8 Gflop/s for double-precision operations. The specification of VEs shows in Table 1.

## *3.2 Performance Measurement of the Original Codes*

Computational time and vector execution performance are measured on the A300-8 for 50 time steps with $1{,}024^3$ grid points. Figure 4 shows the computational time before the optimizations for the two codes. The textures in the bar graph represent the time for some parts of the Fourier transforms and inverse Fourier transforms in each direction. The shaded area indicates the time for all-to-all communication and the transposition in one core.

From `fft_x` in the figure, which denotes the computational time of the x-direction Fourier transforms, it is shown that the computational time of the slab code is longer than those of the other directions, because the second element of the three-dimensional arrays for Fourier coefficients is accessed continuously in the most inner loop for the x-direction FFTs and its access sequences to the memory become non-contiguous.

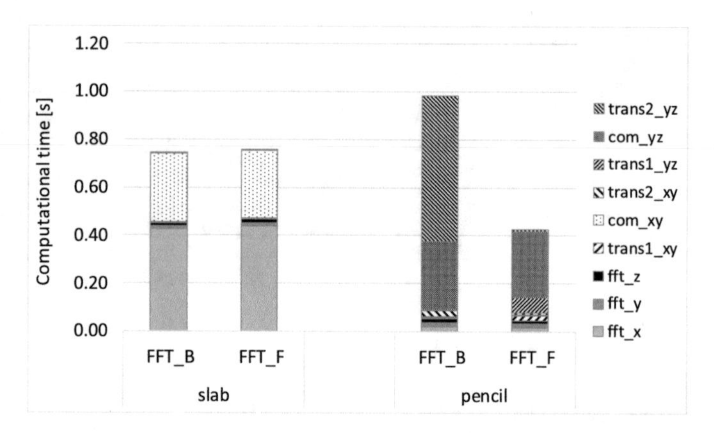


**Fig. 4** Computational time measured on SX-Aurora TSUBASA with $1{,}024^3$ grid points and 64 MPI processes before optimization. FFT − F and FFT − B stand for the 3D-FFT and inverse 3D-FFT, respectively

**Table 2** Average vector length (A. V. length), vector operation ratio (V. Op. ratio), and LLC hit ratio on SX-Aurora TSUBASA for two DNS implementations with 64 MPI processes

| | A. V. length | V. Op. ratio | LLC hit ratio (%) | | CPU port conf. (s) |
|---|---|---|---|---|---|
| | | | Whole | Transposition | |
| slab | 212.8 | 97.9 | 33.4 | 14.6 | 0.22 |
| pencil | 131.4 | 96.9 | 32.5 | 10.0 | 0.05 |

Table 2 shows average vector length, vector operation ratio, LLC hit ratio, and the CPU port conflict time for two codes with 64 MPI processes and $1{,}024^3$ grid points. It is shown the average vector length of the pencil code is shorter than the slab code since the computational area is divided into small portions. The vector operation rate and the LLC hit ratio of two methods is almost the same. However, the LLC hit ratio of the pencil code of the portion of transposition is smaller than that of the slab code. The CPU port conflict time of the slab code is longer than the pencil code. In addition, the total computational time of the slab code is about 1.5 (s) from Fig. 4. The CPU port conflict time accounts for about 15% of the total computational time, because the second element of the three-dimensional array is accessed continuously in the inner-most loop.

### *3.3 Optimization and Its Performance*

First, the slab code is optimized by exchanging the order of the multiple loops in the x-directional Fourier transforms to reduce the CPU port conflict time. In the original

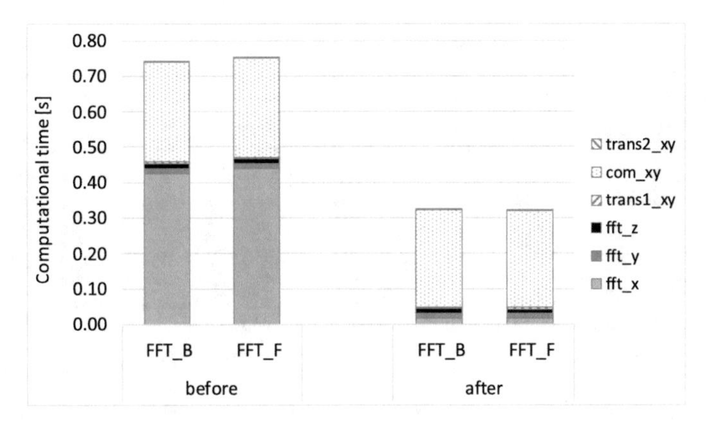


**Fig. 5** Computational time of the optimized the slab code measured on SX-Aurora TSUBASA with $1{,}024^3$ grid points and 64 MPI processes

**Table 3** Average vector length (A. V. length), vector operation ratio (V. Op. ratio), and CPU port conflict time on SX-Aurora TSUBASA for the optimized the slab code with 64 MPI processes

| A. V. length | V. Op. ratio | CPU port conf. (s) |
|---|---|---|
| 219.0 | 98.1 | 0.001 |

code, the first element of the three-dimensional array is accessed continuously to increase the vector length. However, the second element of the three-dimensional variable to be transformed is accessed continuously by changing the order of multiple loops, which could reduce the computational time by reducing the CPU port conflict time.

Figure 5 shows the computational time of the slab code with $1{,}024^3$ grid points and 64 MPI processes after optimization. It is found that the computational time for the x-directional Fourier transform becomes as short as those of the other directions. Table 3 shows the vector performance of the slab code after optimization. By the optimization, the CPU conflict time becomes much shorter than before optimization. Thus, the whole computational time is about 2.2 times shorter than the time by the original code.

For the pencil code, there are copy operations of a 3-dimensional array in a core that make a local transposition of the 3D-FFT, and the part degrades LLC hit ratio. Therefore, loop blocking method is applied to the code to increase the ratio.

Figure 6 shows the computational time of the pencil code by changing block size $ibl$. Table 4 shows the vector performance of the modified pencil code. The smaller $ibl$ is, the higher the LLC hit ratio is as the variables fit into the cache, but the smaller the average vector length as the length of the loop becomes smaller. The calculation time is the shortest at $ibl = 16$, which is 1.6 times shorter than before optimization.

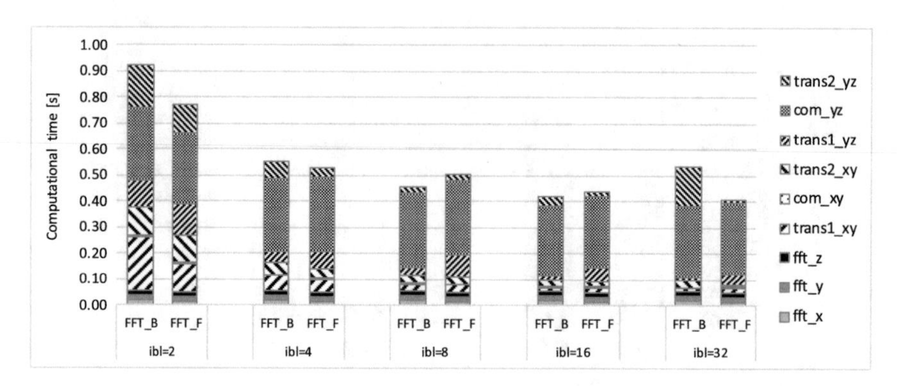


**Fig. 6** Computational time of the optimized the pencil code measured on SX-Aurora TSUBASA with $1{,}024^3$ grid points and 64 MPI processes

**Table 4** Average vector length (A. V. length), vector operation ratio (V. Op. ratio), and LLC hit ratio on SX-Aurora TSUBASA for the optimized the pencil code with 64 MPI processes and blocking size $ibl$

| $ibl$ | A. V. length | V. Op. ratio | LLC hit ratio (%) |
|---|---|---|---|
| 2 | 24.6 | 64.2 | 45.3 |
| 4 | 48.2 | 82.3 | 41.7 |
| 8 | 83.0 | 90.6 | 36.7 |
| 16 | 122.7 | 94.3 | 36.5 |
| 32 | 161.1 | 96.2 | 36.7 |

# 4 Conclusions

In this article, we have optimized the codes by two decompositions; the slab and pencil decompositions on the vector supercomputer SX-Aurora TSUBASA A300-8, and have evaluated its performance. First, the slab code is optimized by exchanging the order of the loop to reduce the CPU port conflict time. After optimization, the CPU port conflict time is reduced from 0.22 to 0.001 s. Second, the pencil code is optimized by loop blocking to improve the LLC hit ratio. The cache hit ratio is improved from 32.5 to 45.3 at best. The computational time of the optimized two codes becomes 2.2 and 1.6 times shorter compared to the time of the original codes, respectively. Peak performances of the slab code are about 1.2% overall and 7.8% for the portion of the calculation excluding transposition, respectively. In addition, those of the pencil code are about 1.0% and 6.5%, respectively.

Reducing the bank conflict time by setting the size of the first element of the array should be appropriately considered. Moreover, the relationship between the cache hit ratio and the block size $ibl$ in the portion of the transposition in one core needs to be clarified.

**Acknowledgements** This work is supported partially by MEXT Next Generation High-Performance Computing Infrastructure and Applications R&D Program, entitled R&D of a Quantum-Annealing-Assisted Next Generation HPC Infrastructure and Its Applications.

## References

1. Orszag, S.A.: Numerical methods for the simulation of turbulence. Phys. Fluids **12**(Supp 1. II), 250–257 (1969). https://doi.org/10.1063/1.1692445
2. Yokokawa, M., Itakura, K., Uno, A., Ishihara, T., Kaneda, Y.: 16.4-Tflops direct numerical simulation of turbulence by a fourier spectral method on the earth simulator. In: SC '02, Proceedings of the 2002 ACM/IEEE Conference on Supercomputing, pp. 1–17 (2002). https://doi.org/10.1109/SC.2002.10052
3. Kaneda, Y., Ishihara, T., Yokokawa, M., Itakura, K., Uno, A.: Energy dissipation rate and energy spectrum in high resolution direct numerical simulations of turbulence in a periodic box. Phys. Fluids **15**(2), L21–L24 (2003). https://doi.org/10.1063/1.1539855
4. Ravikumar, K., Appelahms, D., Yeung, P.K.: GPU acceleration of extreme scale pseudo-spectral simulations of turbulence using asynchronism, SC '19, November 17–22 (2019). https://doi.org/10.1145/3295500.3356209
5. Ishihara, T., Morishita, K., Yokokawa, M., Uno, A., Kaneda, Y.: Energy spectrum in high-resolution direct numerical simulations of turbulence. Phys. Fluids 1 (2016). https://doi.org/10.1063/1.1539855
6. Komatsu, K., et al.: Performance evaluation of a vector supercomputer SX-Aurora TSUBASA. In: SC'18, International Conference for High Performance Computing, Networking, Storage and Analysis, pp. 685–696. Dallas, TX, USA (2018). https://doi.org/10.1109/SC.2018.00057
7. Ishihara, T., Morishita, K., Yokoakwa, M., Enohata, K., Ishii, K.: Accurate parallel algorithm for tracking inertial particles in large-scale direct numerical simulations of turbulence. In: Parallel Computing Technology. Lecture Notes in Computer Science 9251, pp. 522–527. Springer (2015). https://doi.org/10.1007/978-3-319-21909-7_51
8. Yokokawa, M., Morishita, K., Ishihara, T., Uno, A., Kaneda, Y.: Performance of a two-path aliasing free calculation of a spectral DNS code. In: Rodrigues, J.M.F., et al. (eds.) Computational Science—ICCS 2019, pp. 587–595. Springer (2019). https://doi.org/10.1007/978-3-030-22747-0_44
9. Imamura, T., Aoki, M., Yokokawa, M.: Batched 3D-distributed FFT kernels towards practical DNS codes. In: Foster, I., et al. (eds.) Parallel Computing: Technology Trends, pp. 169–178. IOS Press (2020). https://doi.org/10.3233/APC200038
10. Okamoto, N., Matsuzaki, T., Yokokawa, M., Kaneda, Y.: Effect of high-order finite difference discretization of the nonlinear term on turbulence statistics. In: 17th ETC European Turbulence Conference, September 306. Italy, Torino (2019)
11. Yokokawa, M., Nakai, A., Komatsu, K., Watanabe, Y., Masaoka, Y., Isobe, Y.: I/O performance of the SX-Aurora TSUBASA. In: Proceedings of 2020 IEEE International Parallel and Distributed Processing Symposium Workshop (IPDPSW), pp. 27–35 (2020). https://doi.org/10.1109/IPDPSW50202.2020.00014

# Dynamic Load Balancing for Coupled Simulation Methods

**Matthias Meinke, Ansgar Niemöller, Sohel Herff, and Wolfgang Schröder**

**Abstract** A dynamic load balancing technique for simulation methods based on hierarchical Cartesian meshes is presented for two applications in this paper. The first method is a hybrid CFD/CAA solver for the prediction of aeroacoustic noise. In this application, a finite-volume method for the large eddy simulation of the turbulent flow field is coupled to a discontinuous Galerkin method for the solution of the acoustic perturbation equations to predict the generation and propagation of the sound field. The second simulation method predicts a combustion process of a premixed fuel. The turbulent flow field is predicted again by large eddy simulation using the finite-volume method, which is coupled to a level-set solver used for the prediction of the flame surface. In both applications, a joint Cartesian mesh is used for the involved solvers, which allows to efficiently redistribute the computational load using a space filling curve. The results show that the dynamic load balancing can enhance the parallel efficiency even for static meshes. The simulation of the combustion process with a solution adaptvie mesh technique demonstrates the necessity of a dynamic load balancing technique.

## 1 Introduction

The implementation of multiphysics simulation methods on HPC systems is non-trivial due to the necessary exchange of coupling information between the various solvers, usually requiring communication and synchronization of the involved simulation methods. This problem becomes even more severe, when solution adaptive mesh refinement is used, which leads to varying compute loads during a simulation

M. Meinke (✉) · A. Niemöller · S. Herff · W. Schröder
Institute of Aerodynamics, RWTH Aachen University, Wüllnerstrasse 5a, 52062 Aachen, Germany
e-mail: m.meinke@aia.rwth-aachen.de

W. Schröder
JARA Center for Simulation and Data Science, RWTH Aachen University, Seffenter Weg 23, 52074 Aachen, Germany



M. M. Resch et al. (eds.), *Sustained Simulation Performance 2019 and 2020*,
https://doi.org/10.1007/978-3-030-68049-7_5

run. This paper will present some recent results from two applications for which the same dynamic load balancing method detailed in this paper is used.

The first application is connected to the prediction of aeroacoustic noise. The reduction of such noise requires efficient and accurate prediction methods. A well established hybrid computational fluid dynamics/computational aeroacoustics, i.e., CFD/CAA approach [1] couples a large eddy simulation (LES) to compute the unsteady turbulent flow field with a subsequent CAA step, in which, e.g., the acoustic perturbation equations (APE) are solved to predict the acoustic field, i.e., the propagation of the acoustic waves through the inhomogeneous near field to the near far-field. Engine jet noise is, e.g., a major issue, which can be reduced by nozzle shape modifications such as chevrons [2]. The optimization of such nozzle modifications can be performed by this prediction methods implemented on HPC systems [3]. A new fully-coupled direct-hybrid method has been developed, in which the CFD and CAA solutions are conducted in a shared framework, i.e., both solvers are coupled within a single code [4]. This allows the concurrent execution of both solvers with in-memory data exchange of the acoustic source terms.

The simulation of flows with premixed combustion is the second example, in which the concurrent solution of two solution methods is required. In this paper, a coupled LES and level-set (LS) solver is used to predict turbulent swirl flames, which are of major interest in many industrial applications, e.g., gas turbines. The swirling flow stabilizes the combustion process and features a recirculating zone on the centerline of the swirling jet downstream of the fuel injection, which leads to a compact flame with a relatively high rate of fuel burning. Due to the swirling flow a helical flow instability, i.e., a precessing vortex core (PVC), is often observed in combustion chambers. Furthermore, combustion instabilities (CI) may occur as an amplified interaction between acoustic waves in the combustion chamber and the unsteady heat release of the flame. The design and optimization of such combustion processes requires HPC resources, such that the simulation methods have to possess a good parallel efficiency.

By employing a joint hierarchical Cartesian grid, which is partitioned on a coarse level via a space-filling curve (SFC), different solvers using a subset of the joint Cartesian mesh, can be coupled and parallelized efficiently. Solution adaptive meshes can be implemented with dynamic load balancing (DLB), which is beneficial or even necessary to achieve a high parallel efficiency. In the following, a zero-knowledge-type dynamic load balancing scheme for direct-coupled multiphysics applications is presented, which allows to increase the efficiency of complex simulations with non-trivial domain decompositions. It is exemplified using the coupled solvers for the two applications. The novel method comprises a measurement based estimation of computational weights using the compute time on each parallel process and the current distribution of cells of both solvers in all subdomains. The new method operates fully automatically, i.e., it requires no a-priori knowledge or manual assessment to determine suitable weighting parameters. Further, instead of solving the CCP problem, an incremental, diffusive DLB approach is employed to determine a new partitioning of the SFC into parallel subdomains. That is, the change in the partitioning directly depends on the measured compute loads, which allows to alleviate load imbalances more rigorously. In addition to a detailed performance analysis, the applicability is demonstrated for the two applications.

This study has the following structure. First, the numerical methods are introduced in Sect. 2. In Sect. 3, the dynamic load balancing scheme is described. A few simulation results and the analysis of the DLB method are presented in Sect. 4. Finally, conclusions are drawn in Sect. 5.

## 2 Numerical Simulation Methods

In this section the mathematical models used in the two applications and the applied numerical schemes are summarized. The prediction of aeroacoustic noise is based on a two-step approach, since in general two separate simulation tools are used to compute the flow and subsequently, the acoustic field [5]. The direct-hybrid method is used, in which the same simulation framework for the CFD and CAA allows a concurrent simulation of the flow and the acoustic field [4]. A finite-volume method is used for the solution of the Navier-Stokes equations to predict the turbulent flow field. The acoustic field is determined by a discontinuous Galerkin spectral element method for the solution of the APE. For the prediction of the combustion process the Navier-Stokes equations are extended by a progress variable approach, which is used to determine a source term for the heat release in the energy equation. A solution of a level-set equation is used to track the thin reaction zone. All solvers operate on a joint hierarchical Cartesian grid that is partitioned on a coarse level via a space-filling curve. Full subtrees of the grid are then assigned to the parallel subdomain, which facilitates an efficient spatial coupling without any communication overhead.

In the following, the governing equations describing the flow and the acoustic field are introduced in Sect. 2.1. Then, the combustion model and the level-set formulations are described in Sect. 2.3. The numerical methods for the CFD, the CAA and the level-set are given in Sect. 2.4, while Sect. 2.5 focuses on the mesh topology and the partitioning approach of the grid into parallel subdomains. Finally, the coupling approach employed is presented in Sect. 2.6.

### *2.1 Navier-Stokes Equations*

The conservative form of the non-dimensional Navier-Stokes equations for a compressible fluid is given by

$$\frac{\partial \rho}{\partial t} + \nabla \cdot (\rho \boldsymbol{u}) = 0, \tag{1a}$$

$$\frac{\partial \rho \boldsymbol{u}}{\partial t} + \nabla \cdot \left( \rho \boldsymbol{u}\boldsymbol{u} + p\boldsymbol{I} + \frac{\boldsymbol{\tau}}{\mathrm{Re}_0} \right) = 0, \tag{1b}$$

$$\frac{\partial \rho E}{\partial t} + \nabla \cdot \left( (\rho E + p)\boldsymbol{u} + \frac{1}{\mathrm{Re}_0}(\boldsymbol{\tau u} + \boldsymbol{q}) \right) = \mathcal{Q}\bar{\dot{\omega}}_c, \tag{1c}$$

in which the conservative variables comprise the density $\rho$, the velocity vector $\boldsymbol{u}$, and the total specific energy $E = e + \frac{1}{2}\boldsymbol{u}^2$. The source term in the energy equation $\mathcal{Q}\bar{\dot{\omega}}_c$, i.e., the product of the heat realease $\mathcal{Q}$ with the chemical reaction rate $\bar{\dot{\omega}}_c$, in Eq. 1c, is zero for non-reacting flow and is defined in Sect. 2.3 for flows with combustion processes. The internal energy $e$ is linked to the pressure $p$ via the equation of state for an ideal gas to close the equation system $p = (\gamma - 1)\,\rho e$, with the isentropic exponent for air being $\gamma = 1.4$. The non-dimensionalization is based on stagnation quantities denoted by the subscript 0 leading to the Reynolds number $\mathrm{Re}_0 = \rho_0 c_0 L/\mu_0$, with a reference length $L$, the speed of sound $c_0 = \sqrt{\gamma p_0/\rho_0}$, and the dynamic viscosity $\mu_0$. The stress tensor for a Newtonian fluid is given by

$$\boldsymbol{\tau} = -\mu\left[\nabla\boldsymbol{u} + (\nabla\boldsymbol{u})^{\mathsf{T}}\right] + \frac{2}{3}\mu(\nabla\cdot\boldsymbol{u})\boldsymbol{I}. \tag{2}$$

The dynamic viscosity $\mu$ is modeled by Sutherland's law. The heat conduction vector $\boldsymbol{q}$ is computed based on the static temperature $T$ using Fourier's law

$$\boldsymbol{q} = -\frac{k}{\mathrm{Pr}(\gamma - 1)}\nabla T \tag{3}$$

with constant Prandtl number $\mathrm{Pr} = \mu_0 c_p/k_0 = 0.72$, the specific heat at constant pressure $c_p$, and the thermal conductivity $k(T) = \mu(T)$.

## 2.2 Acoustic Perturbation Equations (APE)

In this paper, aeroacoustic noise of compressible flows is analyzed. Therefore, the APE-4 formulation of the acoustic perturbation equations is used [1]

$$\frac{\partial \boldsymbol{u}'}{\partial t} + \nabla\left(\bar{\boldsymbol{u}}\cdot\boldsymbol{u}' + \frac{p'}{\bar{\rho}}\right) = \boldsymbol{q}_m, \tag{4a}$$

$$\frac{\partial p'}{\partial t} + \bar{c}^2\nabla\cdot\left(\bar{\rho}\boldsymbol{u}' + \bar{\boldsymbol{u}}\frac{p'}{\bar{c}^2}\right) = 0, \tag{4b}$$

to determine the acoustic pressure and velocity fluctuations $p'$ and $\boldsymbol{u}'$. Time averaged quantities denoted by an overbar are computed by the compressible flow simulation. The right-hand side source term $\boldsymbol{q}_m = -\boldsymbol{L}' = -(\boldsymbol{\omega}\times\boldsymbol{u})$, consists of the perturbed Lamb vector $\boldsymbol{L}'$ with the vorticity vector $\boldsymbol{\omega} = \nabla\times\boldsymbol{u}$. The source terms are determined by the LES of the turbulent flow field.

## 2.3 Combustion Model

The prediction of the combusting flow is performed based on a progress variable approach, which defines the progress of the chemical reaction and thus the heat release. The progress variable $c$ is computed by the transport equation

$$\frac{\partial \rho c}{\partial t} + \nabla \cdot \left( \rho \boldsymbol{u} c - \frac{1}{\mathrm{Pr}} \rho_\infty^u D \nabla c \right) = \bar{\dot{\omega}}_c, \tag{5}$$

where $\rho_\infty^u$ denotes the density of the unburnt gas in the freestream and the diffusivity $D$ is assumed to be a constant evaluated in the unburnt gas. The chemical reaction rate $\bar{\dot{\omega}}_c$ is given by

$$\bar{\dot{\omega}}_c = Re\, Pr\, \bar{\rho} \frac{\rho_\infty^u}{\rho_\infty^b} R_r \left(1 - c\right) \Psi \left(G(\mathrm{x}, t), \sigma\right), \tag{6}$$

where the quantity $\rho_\infty^b$ is the freestream value of the density in the burnt gas. $\mathcal{Q}$ in Eq. 1c, denotes the heat release of the chemical reaction. The Ferziger and Echeckki model constant $R_r$ reads

$$R_r = \frac{1}{1 - c^0} \left( \frac{1}{1 - c^0} - 1 \right) \frac{(s^u)^2}{D}, \tag{7}$$

where $c^0$ is the reduced inner layer temperature and $s^u$ is the local curvature-corrected flame speed evaluated with respect to the unburnt gas. The function $\Psi\ (G\,(\mathrm{x}, t)\,, \sigma)$ controls the reaction rate profile and depends on the parameter $\sigma$, used to artificially thicken the flame, and the signed distance to the flame front $G$.

The motion of the inner-layer temperature contour of the flame is described by the zero level-set contour, i.e., the $G = G_0 = 0$ contour, of the three-dimensional function $G(\mathrm{x}, t)$. The evolution of the G-field is defined by the $G$-equation [6, 7]

$$\frac{\partial \check{G}}{\partial t} + \left( \tilde{\mathrm{v}} + \frac{\rho_{\infty,u}}{\bar{\rho}} \hat{\check{s}}_{t,u} \check{\mathrm{n}} \right) \cdot \nabla \check{G} = 0, \tag{8}$$

$$\check{\mathrm{n}} = -\frac{1}{|\nabla \check{G}|} \left( \partial \check{G}/\partial x,\ \partial \check{G}/\partial y,\ \partial \check{G}/\partial z \right)^T. \tag{9}$$

For an arbitrary variable $\Theta$ the notation $\check{\Theta}$ indicates a variable defined at the filtered flame front location $\hat{\check{\mathrm{x}}}_f$ such that $\check{G}\ (\mathrm{x}, t) = G_0$, the notation $\hat{\check{\Theta}}$ represents variables, which are filtered by the surface integral over the resolved flame front, $\tilde{\Theta}$ represents Favre filtered, and $\overline{\Theta}$ spatially filtered variables. In Eq. 8, the quantity $\tilde{\mathrm{v}}$ is the local Favre filtered flow velocity, $\rho_{\infty,u}$ the freestream density in the unburnt gas, $\bar{\rho}$ is the spatially filtered density

$$\bar{\rho} = \frac{\rho_{\infty}^{u}}{1 - c\left(\rho_{\infty}^{b} - \rho_{\infty}^{u}\right)/\rho_{\infty}^{b}}, \tag{10}$$

and $\check{n}$ the normal vector at $\hat{\check{x}}_f$ pointing towards the unburnt fuel-air mixture.

More details about the combustion model can be found in [8, 9], where the interaction of acoustic modes with an open turbulent jet flame and the noise sources of lean-premixed flames were analyzed.

## 2.4 Numerical Methods

The turbulent flow field is computed by a large-eddy simulation (LES) on an unstructured hierarchical Cartesian grid using a finite-volume discretization [10]. A monotonic upstream-centered scheme for conservation laws (MUSCL) is employed to determine the state variables on the cell surfaces. The convective terms are approximated by a modified low-dissipation advection upstream splitting method [11], while the viscous terms are computed by central differences. The impact of the subgrid scale terms is modeled via the monotone implicit LES (MILES) approach [12]. Time integration is performed using a five-stage explicit Runge-Kutta method. The boundaries are resolved by a strictly conservative cut-cell approach [13]. Overall, the scheme is second-order accurate in space and time. Further details and applications to various flow problems can be found in [14, 15].

The acoustic perturbation equations are discretized by a nodal discontinuous Galerkin spectral element method (DGSEM) [16] using hexahedral mesh elements. The solution state in each DG element is approximated by Lagrange polynomials of arbitrary degree employing a tensor product ansatz for each spatial direction. The numerical flux on element interfaces is approximated using the local Lax-Friedrichs flux. Non-conforming elements occurring in locally refined Cartesian grids are treated using a mortar element method [17]. An explicit five-stage low-storage Runge-Kutta method [18] is employed for the time integration of the semi-discrete DG operator. A detailed description of the methodology can be found in [4].

The $G$-equation, Eq. 8, is solved by a fifth-order upwind-central scheme [19] for the spatial discretization and the same third-order TVD Runge-Kutta scheme [20]. A high-order constrained reinitialization scheme HCR-2 [21] is used to reinitialize the signed distance property $|\nabla G| = 1$. The computational cost of the level-set solver is minimized by solving the $G$-equation only in a narrow band near the flame front and adapting the level-set grid according to the flame location.

## 2.5 Mesh Topology and Domain Decompositioning

All solvers operate on a hierarchical Cartesian grid in which cells are organized in an octree structure with parent, child, and neighbor relationships [14]. Details of the implementation anf the mesh generation process can be found in [22].

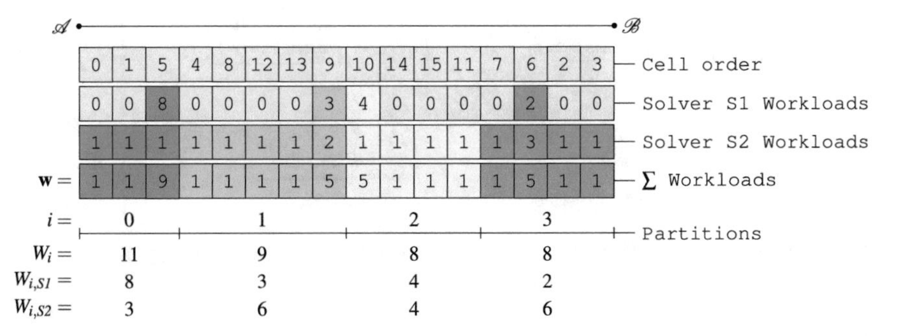


**Fig. 1** SFC linearization, workload of the solvers S1 and S2 and accumulated partition-cell workloads, total workloads $W_i$ of the resulting partitions (solution of the CCP problem, average workload $\overline{W} = 9$) and Solver S1/S2 workload distribution $W_{i,S1/S2}$ among all subdomains. Color indicates domain decompositioning

The partitioning of the computational grid takes place on a coarse partition level, where a Hilbert space-filling curve (SFC) is used to obtain a one-dimensional ordering of all partition cells. Each cell of the grid is assigned the computational workload associated with the solvers using the cell. By traversing the subtrees of the grid, the accumulated workload for each coarse partition cell is then determined. Thus, the parallelization is reduced to a so-called chains-on-chains partitioning (CCP) problem [23].

That is, a chain of weighted computational tasks is contiguously mapped onto a chain of parallel processes. Consequently, a domain decomposition can be obtained by splitting the one-dimensional workload distribution into partitions of similar total workload. The approach is illustrated in Fig. 1 using the linearization of cells along the Hilbert SFC connecting point $\mathscr{A}$ and $\mathscr{B}$. The list of workloads $\mathbf{w}$ comprises the partition-cell workloads $w_k$. Solving the CCP problem, for which $\overline{W} = 9$ is the average workload per process, the partitioning into subdomains is obtained. The domain offsets $o_j$ are given by the splitting positions, which correspond to the first partition cell of each domain. With $W_i$ the workload of the $i$th domain, the efficiency of the partitioning is given by the maximum domain workload. Accordingly, the partition quality [24] is assessed by $P = \frac{\overline{W}}{\max_i W_i}$, which is $P = \frac{9}{11} \approx 82\%$ in the given example. Each partition, i.e., a continuous range of partition cells with the corresponding subtrees of the grid, is allotted to a dedicated parallel process. To prevent coarse-grained partitionings, the partition level can locally be shifted to a higher refinement level, if the partition cell workload becomes too large.

## 2.6 Solver Coupling

A joint hierarchical Cartesian grid is used for all solvers, see e.g. [4]. Refinement constraints for each physical system are taken into account during the grid generation

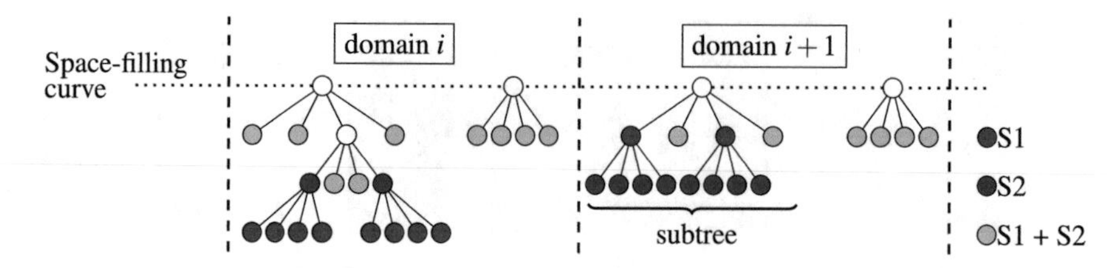


**Fig. 2** Domain decomposition of a quadtree grid with cells used for solver S1, S2, or both

process. All cells are tagged according to their use by either solver. Due to different domain sizes, regions can exist in which only a single solver is active. The domain decomposition is performed on the coarse partition level via the Hilbert SFC, taking into account the computational workloads of the cells.

In the domain decomposition sketc.hed in Fig. 2, with in total 28 and 22 active cells for the two domains, the local share of solver S1 cells is $\frac{17}{28} \approx 61\%$ and $\frac{8}{22} \approx 36\%$. This partitioning strategy facilitates an efficient spatial coupling, since all solver S1 and S1 cells contained in the same volume of the three-dimensional domain are assigned to the same process allowing in-memory exchange of the the coupling terms. In case of the aeroacoustic simulation, the acoustic sources have to be passed from the CFD to the CAA solver, while for the combustion simulation, the location of the flame surface has to be passed from the level-set solver to the CFD solver and the flow field velocity has to be sent from the CFD to the level-set solver. Spatial interpolation from the solver S1 to the solver S2 cells is performed by local Galerkin projection [25], which can handle arbitrary spatial mappings while requiring only data that is locally available.

In this study, the same time step is used for all solvers. Thus, temporal interpolation of the source term data in the CFD/CAA simulation or the flame speed in the combustion model is not required. The parallel coupled algorithm consecutively and alternately advances the Runge-Kutta time integration stages of both solvers. After calculating the coupling terms between two solvers, the individual solutions are consecutively advanced by one Runge-Kutta stage until the time step is completed. This interleaved execution pattern prevents any overhead due to differing load compositions and allows to overlap communication with computations. A detailed description of the coupling approach can be found in [4, 25].

## 3 Dynamic Load Balancing Method

The goal of dynamic load balancing is to maximize the parallel efficiency of a computation by redistributing the workload among processes such that the maximum available computing power is used [26]. Load imbalances and the resulting negative impact on the performance can be quantified by different metrics. Here, the imbalance percentage $I_{\%}$ is used [27], which is defines the severity of an imbalance $I_{\%} = (t_{max} -$

$t_{avg}) \cdot N/(t_{max} \cdot (N-1))$. The quantities $t_{max}$ and $t_{avg}$ represent the maximum and average time to process a given section of code and $N$ is the number of parallel processes. The imbalance percentage is defined by the ratio of the total amount of time wasted $(t_{max} - t_{avg}) \cdot N$ to the amount of parallel resources $t_{max} \cdot (N-1)$. A value of $I_{\%} = 0\%$ corresponds to a perfectly balanced load distribution, while $I_{\%} = 100\%$ is associated with a code section executed only on a single process. Thus, the metric indicates the amount of wasted resources. The potential run time savings are expressed by the imbalance time $I_t = t_{max} - t_{avg}$ assuming that perfect load balance can be achieved [27]. Furthermore, the allocation-time impact is estimated by $I_T = N \cdot I_t$, which is an upper bound on the total amount of wasted resources [28].

Different strategies exist for DLB in scientific computing applications [26, 29]. Dynamic load balancing of large-scale coupled multiphysics simulations is inherently complex and poses various challenges. For instance, multi-stage calculations and interleaved computations with inherent communication barriers may prevent load balancing [30, 31]. The direct-hybrid CFD/CAA method is an example for such a coupled multiphysics simulation and even for static load distributions, a parallel computation might exhibit load imbalances.

Without load balancing, the parallel efficiency of the overall computation is determined by the initial domain decomposition, which is based on estimates regarding the different computational costs for the involved solver cells. These estimates can be determined by relating measured run times obtained from simulations involving only a single solver. This approach, however, usually does not take into account the idle times of all parallel subdomains or, for example, the various boundary conditions that might be computationally more expensive. Consequently, the predicted load can deviate significantly from the actual load on each parallel subdomain during the coupled simulation [32].

This performance impact is exacerbated when considering large-scale simulations with relatively small domain workloads or when using heterogeneous computing hardware [33]. Then, a DLB method is required that automatically redistributes the computational workload to reduce imbalances during the simulation. For coupled solvers, load balancing requires the redistribution of mesh cells implicitly ensuring locality of the coupling terms to be exchanged between the solvers.

In the following, a framework for coupled simulations is extended by a DLB algorithm that satisfies the stated constraints. Based on measurements of the computing time on each parallel process and the current distribution of cells of all solvers in all subdomains, new computational weights are computed (Sect. 3.1). These weights are then used to determine a new domain decomposition (Sect. 3.2).

### *3.1 Computational Weights of Solver Cells*

Key to partitioning and load balancing of coupled multiphysics simulations is the estimation of computational effort for the solver algorithms involved in the simulation [34]. The standard approach for single-solver frameworks aims at distributing cells evenly among all parallel subdomains, assuming a homogeneous computing environment. This poses, however, difficulties when multiple solvers with varying computational costs exist. This problem can be reduced by using a-priori determined computational weights for the CFD, CAA and level-set solver cells. For complex applications, however, it may be difficult to obtain exact computational weights, since, e.g., the distribution of cells with more or less complex boundary conditions is not known in general.

The DLB algorithm proposed in this study estimates computational weights based on measurements of the computing time on each parallel subdomain. Furthermore, the current distribution of solver cells among all subdomains is incorporated. During the simulation, the collective compute time of all solvers is measured locally for each time step. By employing the interleaved execution pattern of the Runge-Kutta substeps and using non-blocking communication, only the total workload independent of the local workload composition needs to be balanced among all domains. This implies that blocking point-to-point communication cannot used in the solvers. On each subdomain $i$, the average compute time $r_i$ is determined as the 25% truncated mean to filter out program-external influences such as system noise [28]. When the global average compute time among all $N$ parallel processes is given by

$$\overline{r} = \frac{1}{N}\sum_{i=0}^{N-1} r_i, \tag{11}$$

the local computational load $l_i$ is determined as $l_i = r_i/\overline{r}$.

Subsequently, the computational weights $\boldsymbol{c}$ for the different load types can be estimated by solving the least squares problem $\boldsymbol{Ac} = \boldsymbol{l}$, with the right-hand side given by the load vector $\boldsymbol{l} = (l_0, l_1, \ldots, l_{N-1})^{\mathsf{T}}$ and the left-hand side matrix $\boldsymbol{A}$ representing the current workload distribution among all subdomains discussed in Fig 1. This assumes that on average the load can be expressed as a linear combination of the individual workload contributions. The linear least-squares problem with unique minimum-norm solution can be written as

$$||\boldsymbol{Ac} - \boldsymbol{l}||_2 = \min_{\boldsymbol{v}} ||\boldsymbol{Av} - \boldsymbol{l}||_2 \quad \text{and} \quad ||\boldsymbol{v}||_2 \text{ is minimal.} \tag{12}$$

In the present work, the DGELSD routine of LAPACK [35] is used to solve the least-squares problem given in Eq. 12.

## 3.2 Subdomain Decompositioning

A domain decomposition based on SFCs reduces to solving the resulting CCP problem. A straightforward extension for dynamic load balancing is given by estimating the computational weights for different load types during the simulation to determine new workloads as an input to the 1D partitioning problem. Since local workload variations are not captured, the partitioning may not be optimal in terms of performance. Therefore, to rigorously alleviate load imbalances, a novel procedure is proposed to determine domain offsets for the SFC-based partitioning.

In the following repartitioning steps, the individual domain offsets are iteratively refined based on the measured load imbalance and the computed computational weights. The number of iterations for the DLB strategy to converge is problem dependent. Thus, for problems exhibiting a static imbalance it is rational to limit the number of iterations such that after an initial balancing phase the computation can continue with the best partitioning determined by then. As a starting point of the DLB approach, the cumulative load imbalances given by

$$s_j = \sum_{i=0}^{j-1} l_i - 1.0 \quad \forall j \in \{1, \ldots, N\}, \quad \text{with} \quad s_0 = 0, \tag{13}$$

are determined. Each value quantifies for the corresponding domain offset $o_j$ the overall load imbalance of domains left and right of the splitting position. Thus, an optimal local position requires shifting the offset $o_j$ along the SFC to minimize the cumulative imbalance. The general assumption of this approach is that by optimizing each offset individually global load balance can be obtained.

The approach is depicted in Fig. 3, with a list of partition cells split into four partitions. With the computational load $l_i$ of each parallel subdomain, the cumulative imbalances $s_j$ are computed according to Eq. 13. This implies, considering the value of $s_2 = 0.45$ in the example, that the compute load of the first two processes is significantly higher than that of the two remaining ones. That is, the accumulated workload of process 0 and 1 is 45% above average. Therefore, by shifting the offset to the left, workload is moved from overloaded to underloaded processes. Accordingly, global load imbalances can be reduced by individually shifting each domain offset in the direction given by

$$d_j = -\operatorname{sign}\left(s_j\right). \tag{14}$$

The load share $\widetilde{w}_k$ of the partition cell $k$ on domain $i$ is computed as

$$\widetilde{w}_k = l_i \cdot \overline{w}_k, \quad \text{with} \quad \overline{w}_k = \frac{w_k}{W_i}, \tag{15}$$

which can be interpreted as an allocation of the load $l_i$ onto the local partition cells. By counterbalancing the cumulative imbalance $s_j$ with the traversed partition cell load shares

$$\begin{aligned} s_j^k &= s_j^{k-1} + d_j \cdot f_{penalty} \cdot \widetilde{w}_{m(k)} \quad \text{for} \quad k \geq 1, \\ \text{with} \quad m(k) &= o_j + d_j k - \frac{1}{2}\left(d_j + 1\right) \quad \text{and} \quad s_j^0 = s_j, \end{aligned} \tag{16}$$

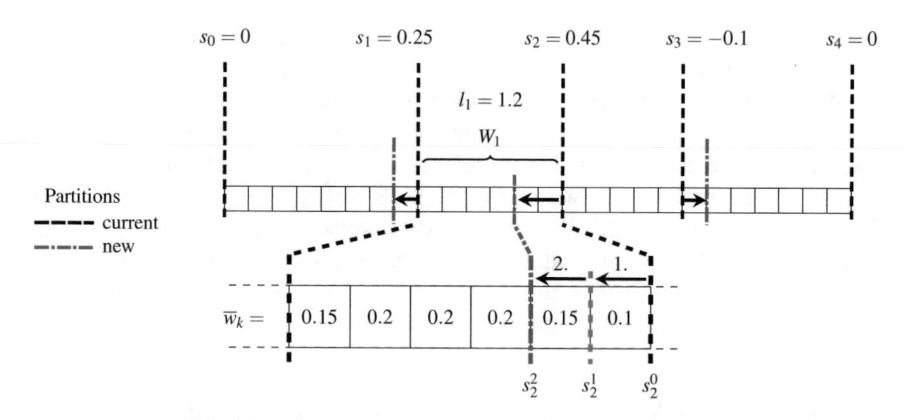


**Fig. 3** Basic concept of the DLB approach, with a list of partition cells split into four parallel subdomains. Combining the computational loads $l_i$, the cumulative imbalance $s_j$ at each domain offset is determined. The new partitioning is obtained by shifting each offset until according to Eq. 16, the imbalance is predicted to be counterbalanced. The schematic illustrates this procedure for the second domain offset, given the corresponding partition cell workload shares and a measured overload of 20% on the second domain

the necessary displacement of each offset $o_j$ can be assessed by the sequence index $k$ for which $s_j^k \approx 0$ holds. That is, the cumulative load deviation $s_j^0$ is balanced by the accumulated load shares $\widetilde{w}_{m(k)}$, which are used to assess the amount of transferred workload, with $m(k)$ the partition cell index to the sequence index $k$. The additional penalization factor $f_{penalty} \geq 1$ allows to limit the displacements and prevent overshooting. Thus, the DLB algorithm refines the partitioning during the simulation. This approach is illustrated in Fig. 3 and Eq. 17 for the second domain offset $o_2$, using a penalty factor of $f_{penalty} = 1.25$. According to Eq. 16, the initial cumulative imbalance of $s_2 = 0.45$ is minimized by shifting the offset by two partition cells, which yields the new domain offset

$$\begin{aligned} s_2^0 := 0.45 \quad &\Rightarrow \quad s_2^1 := 0.45 - 1.25 \cdot 1.2 \cdot 0.1 = 0.3 \\ &\Rightarrow \quad s_2^2 := 0.3 - 1.25 \cdot 1.2 \cdot 0.15 = 0.075. \end{aligned} \tag{17}$$

When proceeding with the computation, the overload on process 0 and 1 will be reduced, since the workload of the two partition cells is transferred to the previously underloaded processes. Furthermore, by introducing performance factors estimating the processing speed on differing compute nodes, the approach can be extended to target al.so computations on heterogeneous computing hardware. In general, this DLB approach is suitable for applications with static or dynamic load imbalances. In the latter case, however, the cost of frequent redistributions needs to be balanced against the potential performance gains, see Sect. 4.4.

# 4 Results

## *4.1 Jet Noise Prediction*

The CFD/CAA simulation of a turbulent jet is performed to demonstrate the effectiveness of the presented dynamic load balancing scheme. In the following, the cold jet flow exhausting from the NASA Glenn nozzle SMC006 [2] is considered. This chevron nozzle shown in Fig. 4 consists of six symmetric lobes with penetration angle 18.2° into the jet stream. The flow conditions are chosen according to the test point 7 of Tanna [36], i.e., the jet temperature related to the ambient temperature $T_j/T_\infty = 0.84$ and an acoustic Mach number based on the jet exit velocity of $Ma_{ac} = U_j/c_\infty = 0.9$. The nozzle exit diameter based Reynolds number is $\mathrm{Re}_D =$ 400,000. More details about the numerical setup for the CFD simulation can be found in [37]. A mesh with about 170 million grid cells is used, where the highest grid resolution covers the acoustic source region. It has an extent of $10D$ in the downstream and $5.5D$ in the sideline direction. In the CAA simulation, cells in the source region have a four-times lower spatial resolution. The DG scheme uses a polynomial degree of $p = 3$ yielding about 350 million degrees of freedom. A snapshot of the predicted acoustic field is shown in Fig. 5. The power spectral density (PSD) of the acoustic pressure fluctuations is evaluated at the radial distance $r/D = 7.5$ for the two axial positions $x/D = 7.5$ and $x/D = 12.5$. Using a sampling interval of $\Delta t_s c_0/D = 0.0025$ the sampling time being started at the simulation time $t_a c_0/D = 24$ is $T_{s,a} c_0/D = 164$. In Fig. 6 the computed near far-field power spectra plotted over the Strouhal number $\mathrm{St} = fD/U_j$ are compared with results from a standard hybrid scheme [37]. Overall, the results of the direct-hybrid method closely agree with the reference data, especially in the low frequency range $\mathrm{St} < 0.4$. In the high frequency range for the axial position $x/D = 7.5$, the decay of the power spectral density is less pronounced in the direct-hybrid method, since the used CAA setup provides a higher spatial and temporal resolution than the reference data, which allows to better resolve the high-frequency noise components propagating normal to the jet axis.

**Fig. 4** Geometry of the NASA Glenn nozzle SMC006 [2]

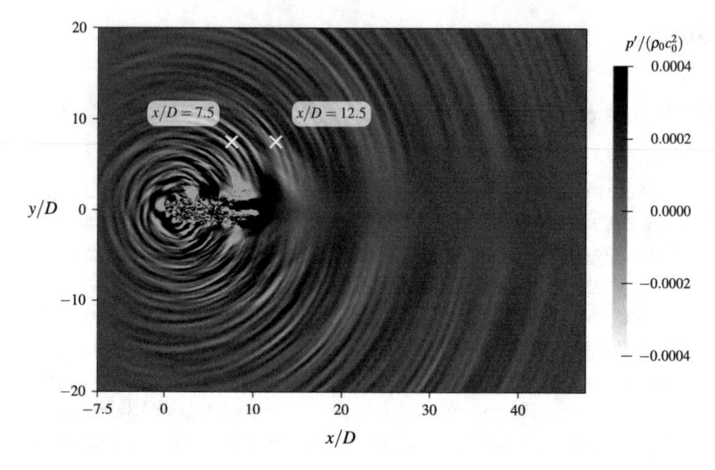


**Fig. 5** Noise prediction for a jet emanating from a chevron nozzle SMC006. Acoustic pressure field in the $x$-$y$-plane and observer locations at a radial distance of $r/D = 7.5$

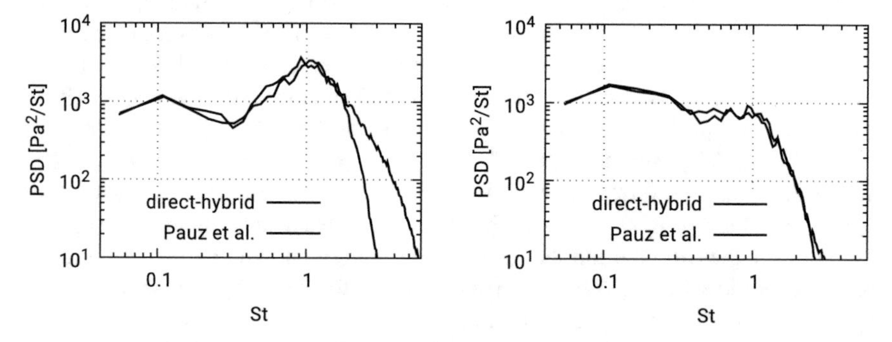


**Fig. 6** Power spectral density (PSD) of the acoustic pressure in the near far-field at the radial distance $r/D = 7.5$ for $x/D = 7.5$ (left) and $x/D = 12.5$ (right) with reference data computed by a standard hybrid scheme from Pauz et al. [37]

## 4.2 Simulation of a Premixed Combustion Process

The burner geometry is depicted in Fig. 7. It corresponds to the experimental burner configuration by Moeck et al. [38].[1]

The geometry includes a part of the burner plenum with a converging nozzle, a static radial swirler, the injector, and the combustion chamber. A mixture of methane and air with a fuel-lean equivalence ratio of $\phi = 0.67$ enters the burner plenum. Downstream of a converging nozzle an angular momentum is generated by the radial swirler through 9 blades. The current study uses the manufactured swirler geometry to allow a comparison with experimental findings. The resulting swirling jet has a swirl number, defined by the ratio of the axial fluxes of the azimuthal momentum

[1]The geometry of the burner has been provided by EM2C.

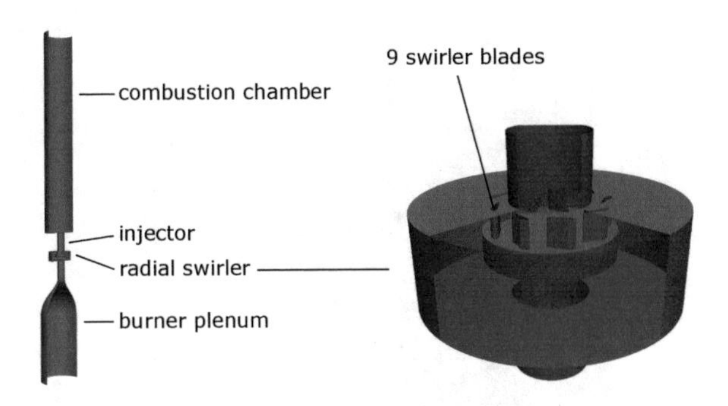


**Fig. 7** Geometry of the burner (left) and the radial swirler (right)

and the axial momentum [39]

$$S = \frac{\int_0^R \rho v_{ax} v_\theta 2\pi r^2 dr}{R \int_0^R \rho v_{ax}^2 2\pi r dr}, \tag{18}$$

of $S = 0.73$. The Reynolds number based on the exit diameter of the injector $D = 0.012$mm is $Re_D = 8800$.

The swirl flame is symmetrically confined by a combustion chamber with a diameter of $4.2D$ and a length of $25D$. Since no detailed information of the wall temperature distribution from the experimental burner exists, an adiabatic no-slip boundary condition is applied to the burner walls.

The domain is extended downstream of the combustion chamber by an additional volume to ensure the correct acoustic impedance downstream of the injector, which is essential for the computation of confined turbulent flames. Besides a parabolic velocity profile being imposed at the inlet of the plenum, non-reflecting Navier-Stokes characteristic boundary conditions (NSCBC) [40], which reduce acoustic reflections at the boundaries, using a linear relaxation coefficient as suggested in [41] are imposed on the inlet and outlet boundaries. On all other boundaries, a slip wall boundary condition is applied.

The highest mesh resolution is used around the swirler and in the flame region, where a cell spacing of $\Delta x = 0.0115D$ is chosen. This leads to approximately $200 \times 10^6$ cells for the FV solver. The swirler device is resolved with approximately 25 cells along the swirler blades. Since the LS solver has band cells on the highest resolution only in the flame region, its adaptive mesh consists of fewer cells, i.e., there are approximately $7 \times 10^6$ LS band cells.

In Fig. 8b the spectrum of the axial velocity component is shown at $(r, y) = (\frac{1}{3}D, \frac{1}{6}D)$. The two peaks can be associated with vortex breakdown modes. The peak at approximately 880 Hz corresponds to the asymmetric PVC mode and the peak at approximately 1800 Hz is caused by a symmetric helical instability mode. The frequency of the PVC mode is in excellent agreement with the PVC frequency 880 Hz measured in the experiments.

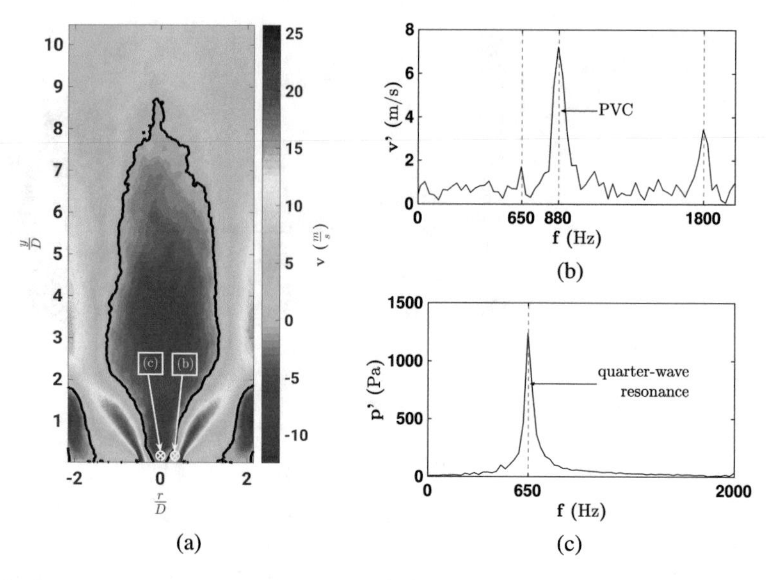


**Fig. 8** Contours of the mean axial velocity component downstream of the injection tube (**a**), spectrum of the fluctuations of the axial velocity component at $(r, y) = (\frac{1}{3}D, \frac{1}{6}D)$ (**b**), and spectrum of the pressure fluctuations at $(r, y) = (0D, \frac{1}{6}D)$ (**c**). The locations of the signals from **b** and **c** are defined in **a** by the white symbols

As mentioned above, the simulations show an acoustic mode at the quarter-wave resonance frequency of the combustion chamber. The spectrum of the pressure in the combustion chamber is shown in Fig. 8c. The frequency of this quarter-wave mode is approximately 650Hz in the simulation, which is higher than the frequency 540Hz observed in the experimental burner. This discrepancy is caused by the assumption of adiabatic walls in the LES.

## 4.3 Dynamic Load Balancing

The dynamic load balancing scheme is first employed in the direct-hybrid simulation of the turbulent jet. The CFD solver uses highly refined cells in the acoustic source term region to accurately predict the turbulent motion of the flow. Farther away, the resolution requirements are lowered such that the grid cells become gradually coarser. In total, about 57.3 million cells are used in the CFD computation.

The CAA domain encloses the CFD domain with an extent of $41D$ and $30D$ in the downstream and sideline direction. The level difference in the acoustic source region between CAA and CFD cells is two, i.e., CAA cells have a four-times lower spatial resolution. In the near far field grid cells with a uniform size are employed. Overall, the CAA simulation uses about 3.71 million cells. A polynomial degree of $p = 3$

is chosen for the DG solution representation, yielding approximately 237.4 million degrees of freedom. The initial mesh partitioning is based on an a-priori estimated computational weight ratio between CAA and CFD cells. A CAA cell with 64 degrees of freedom is weighted by a factor of $w_{CAA} = 8.5$ in relation to a CFD cell. This ratio was determined by relating measured run times for CFD and CAA simulations, which were obtained from standalone simulations of the considered case without coupling of the solvers on 3072 cores.

A strong scaling of the described direct-hybrid simulation is carried out from 192 to 6144 compute cores. The performance experiments were conducted using the Cray XC 40 "Hazel Hen" system of the High Performance Computing Center Stuttgart, Germany (HLRS). All compute time measurements were repeated at least three times from which the result with the minimum run time per time step was selected for comparison.

In the following, Sect. 4.3.1 gives a detailed performance analysis for the simulations on 3072 compute cores. The results of the strong scaling experiment are presented in Sect. 4.3.2.

### 4.3.1 Parallel Efficiency Analysis

First, the run time composition for the jet simulation on 3072 compute cores without and with load balancing is shown in Fig. 9. The average wall time per simulation time step is split into computation and communication among all parallel subdomains. In addition to the CFD and CAA compute loads, the cost for the spatial interpolation of the acoustic source terms is presented. The CAA/CFD communication times correspond to the average time waiting for the completion of an inter-rank data exchange. For clarity, the compute cores are sorted by the CFD load in decreasing order. The reference simulation evidences that the initial domain partitioning without load balancing is suboptimal regarding the workload distribution. In total, the communication overhead accounts for 24% of the total accumulated run time. This value is reduced to 11% after applying DLB, which more evenly distributes the compute load among the parallel subdomains. With a limited number of 20 DLB iterations, the imbalance percentage $I_{\%}$ is diminished from 23% to 10%, corresponding to a significant performance improvement with run time savings of 12.5% even though an already quite good initial mesh partitioning was used.

### 4.3.2 Strong Scaling Experiment

The total computational resources required per simulation time step for a strong scaling experiment from 192 to 6144 compute cores without and with DLB are depicted in Fig. 10. Additionally, the average idle times are shown for each configuration. For lower numbers of compute cores, the parallel computations exhibit only small imbalances. Due to the low degree of parallelism, each domain is assigned a huge compute load that keeps imbalances caused by, e.g., inaccurate computational

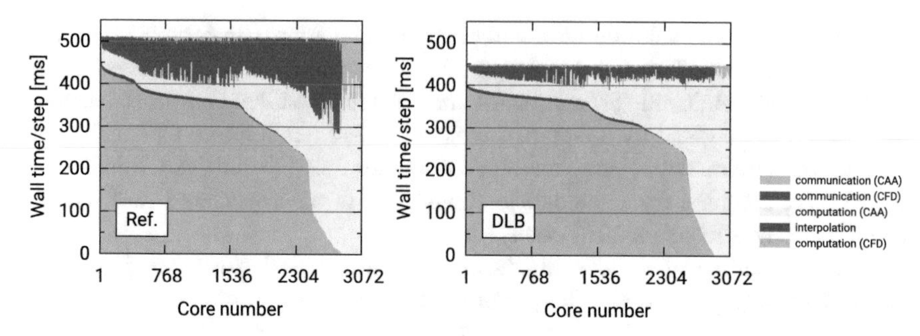


**Fig. 9** Run time distribution for the jet simulation on 3072 cores without (left) and with load balancing (right). Compute cores are sorted by the CFD compute load in decreasing order

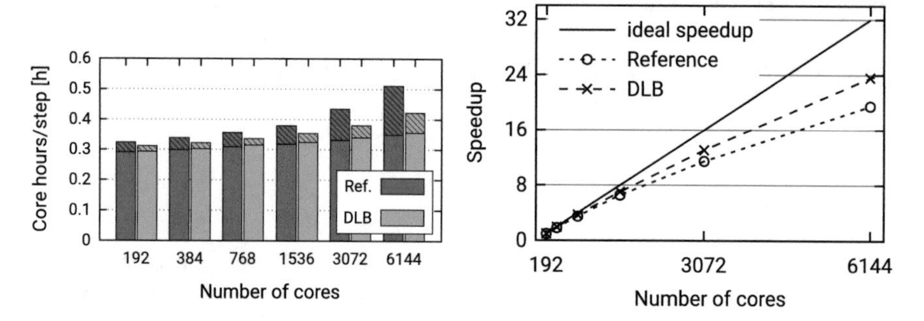


**Fig. 10** Results for a turbulent jet simulation without (Ref.) and with dynamic load balancing (DLB) from 192 to 6144 compute cores. Total computational resources required per simulation time step. Hatched parts correspond to the average idle time (left) and strong scaling test using the compute time with 192 cores as a reference (right)

weights, relatively small. Nevertheless, for DLB small improvements regarding the overall parallel performance are observed. The load imbalances in the reference simulations start with an imbalance percentage of $I_{\%} = 9.6\%$ for 192 cores and increase to $I_{\%} = 31.2\%$ for 6144 cores.

Simultaneously, the total amount of computational resources required to perform the computations in the simulation rises by nearly 20%, due to decreasing local problem sizes. With load balancing, the imbalances are reduced to about $I_{\%} = 2.6\%$ and $I_{\%} = 14.6\%$ for the simulations on 192 and 6144 compute cores. Thus, the benefit grows with the degree of parallelism, saving 17.5% of computational resources when using 6144 cores.

The speedup of the strong scaling experiment is depicted in Fig. 10, with the DLB measurement for 192 cores as a reference for the scalability analysis. The parallel efficiency of the reference simulation declines to about 61% during the scaling, whereas with load balancing an efficiency of 74% is obtained with the 32-fold increase in the number of compute cores.

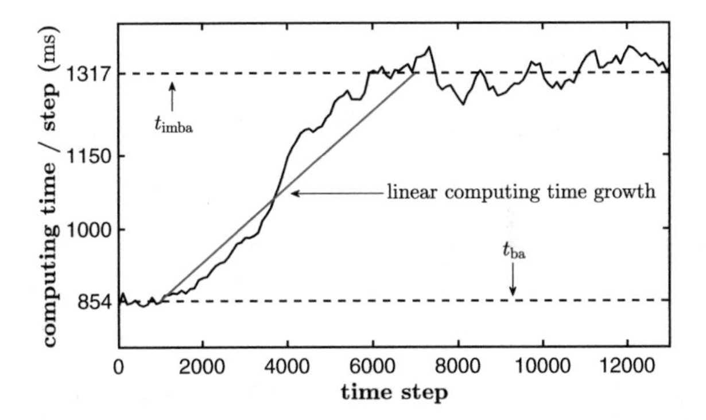


**Fig. 11** Required computing time per time step for the turbulent swirl burner computation on 172,296 compute cores. The DLB method is applied during the first 1000 time steps. The average computing times for the balanced and the imbalanced workload distributions are denoted by the dashed lines $t_{ba}$ and $t_{imba}$. The red line is a linear fit for the growth of the computing time after the DLB method is no longer applied

## *4.4 Mesh Adaptation and Dynamic Load Balancing*

The following analyses were performed for the simulation of the combustion process explained in Sect. 4.2. The coupled CFD/LS simulation uses a static grid for the CFD solver and a solution adaptive mesh that is dynamically adjusted to the flame front for the LS solver. Therefore, an initially well balanced domain partitioning will become increasingly imbalanced due to the change of the computational workload caused by the flame motion. The state of the worst case partitioning is reached when the flame front moves into a subdomain which initially had no LS band cells, i.e., the whole subdomain is fully refined for the LS mesh. This worst state can move from one subdomain to another with the flame movement, i.e., the rotation of the PVC.

To demonstrate the efficiency of the DLB method for a highly parallel computation the following results are based on a large-scale computation on 172,296 compute cores using a joint mesh with 3.5 billion FV cells and an average of roughly $50 \times 10^6$ LS cells and a smallest cell spacing of $\frac{\Delta x}{2} = 0.00575D$.

The required computing time per time step is illustrated in Fig. 11. In the beginning of the computation, i.e., during the first 1000 time steps, the DLB method is applied such that a well-balanced partitioning with an approximately constant computing time per time step is achieved. Then DLB is switched off, that is after 1000 time steps, the computation uses a static partitioning which was determined at the last DLB step such that the computation exhibits an increasing load imbalance.

The required total computing time per time step using the well-balanced partitioning is $t_{\text{ba}} \approx 0.85$s. The computation gradually slows down without DLB due to the increasing workload imbalance, until the required time per time step reaches a plateau with an average time of $t_{\text{imba}} \approx 1.32$s after $\Delta t_{\text{imba}} = 6000$ time steps. That is, the required computing time for one time step is increased by about 55%.

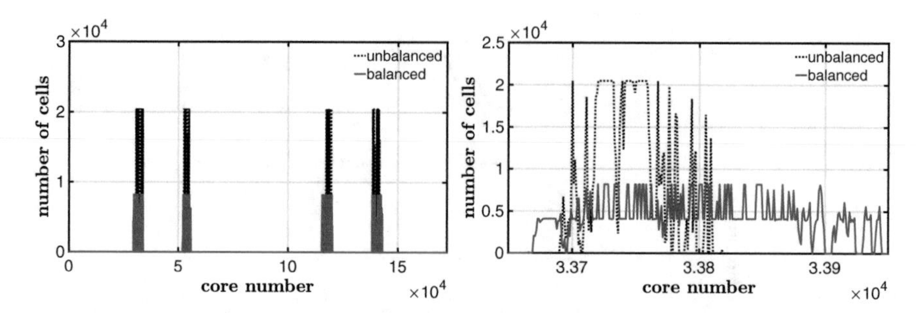


**Fig. 12** Distribution of level-set cells among all CPU cores (left), a set of CPU cores in the core range 33,650–33,950 (right). The black dotted line denotes the distribution in an unbalanced configuration and the red line in a balanced configuration

The load imbalance is caused by the clustering of LS cells, which is further illustrated in Fig. 12 by the distribution among all compute cores (Fig. 12a) and for the localized core range [33650, 33950] (Fig. 12b). With the DLB method the level-set band cells are distributed such that their maximum number is below 9000 cells among all cores. Without DLB these cells cluster on only a few compute cores. The maximum number of level-set band cells on one core is increased up to 20,000 cells. This can be associated with the required higher computing time of the LS solver. The times per time step depicted above exclude the time required to perform the DLB, which is approximately $t_{\mathrm{DLB}} = 46$s per DLB step on 172,296 compute cores.

To determine the optimum DLB interval the potential performance gains have to be balanced against the time required to perform the DLB. Assuming a constant growth of imbalance, which is a reasonable assumption for the current turbulent swirl flame due to the PVC, it is straightforward to determine the total average time required per time step $t_{\mathrm{total}}$ including the time required to perform the DLB method

$$t_{\mathrm{total}} = t_{\mathrm{ba}} + \frac{(t_{\mathrm{imba}} - t_{\mathrm{ba}})}{2 * \Delta t_{\mathrm{imba}}} \Delta t_{\mathrm{DLB}} + \frac{t_{\mathrm{DLB}}}{\Delta t_{\mathrm{DLB}}}. \tag{19}$$

The quantity $t_{\mathrm{ba}}$ is the computing time using a well-balanced partitioning, $t_{\mathrm{imba}}$ is the required computing time for the worst case partitioning, $\Delta t_{\mathrm{imba}}$ is the number of time steps in which the workload imbalance grows, $t_{\mathrm{DLB}}$ is the time required to perform a DLB step, and $\Delta t_{\mathrm{DLB}}$ is the time interval over which the DLB method is applied.

In Fig. 13, the estimated total computing time per step, which is approximated by Eq. 19, is shown as a function of the interval $\Delta t_{\mathrm{DLB}}$ in which the DLB method is applied.

The optimum interval is determined as $\Delta t_{\mathrm{DLB,opt}} = 1092$ time steps and leads to an average required time per time step of $t_{\mathrm{total,opt}} = 0.9383$s. Therefore, the overall computational cost of this computation can be reduced by approximately 30% through the DLB method.

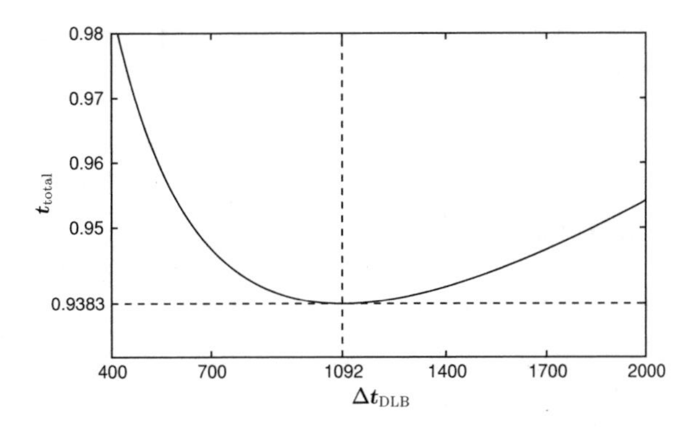


**Fig. 13** Estimated total average computing time required per time step $t_{total}$ as a function of the DLB interval $\Delta t_{DLB}$

## 5 Conclusions

To ensure high parallel efficiency for large-scale coupled multiphysics simulations is challenging since even minor load imbalances can severely impact the overall performance of large scale parallel computations. A dynamic load balancing scheme for direct-coupled multiphysics applications was presented, which increases the efficiency of complex simulations with non-trivial domain decompositions. The novel method is exemplarily applied in two problems, i.e., in a coupled CFD/CAA simulation for aeroacoustic problems and a CFD/LS solver for the prediction of flows with combustion. In both cases, distinct numerical methods were used with varying computational costs per mesh cell. The partitioning is obtained via a Hilbert space-filling curve on a coarse level of the hierarchical mesh. An incremental, diffusive DLB algorithm based on the SFC partitioning was presented, which allows individual adjustments of the domain decomposition.

A three-dimensional turbulent jet CFD/CAA simulation demonstrates the effectiveness of the DLB scheme for large-scale coupled multiphysics simulations. A detailed performance analysis showed the necessity of a DLB method to directly target load imbalances, which are, e.g., caused by the individual efficiency of the computation depending on the local workload composition and the scalability of the individual solvers. Furthermore, a strong scaling experiment showed performance improvements at growing degrees of parallelism, when a-priori estimated computational weights for the initial partitioning are used. The application of the DLB in a simulation of a combustion process involving dynamically changing meshes for the level-set solver, showed the DLB to be necessary to avoid a severe parallel performance degradation due to the dynamically changing level-set cells in the individual subdomains.

**Acknowledgements** This study was funded by the Deutsche Forschungsgemeinschaft (DFG, German Research Foundation), DFG project number 335858412 and 247310774. The authors gratefully acknowledge the Gauss Centre for Supercomputing (GCS) for providing computing time for a GCS Large-Scale Project on the GCS share of the supercomputer "Hazel Hen" at HLRS Stuttgart. GCS is the alliance of the three national supercomputing centres HLRS (Universität Stuttgart), JSC

(Forschungszentrum Jülich), and LRZ (Bayerische Akademie der Wissenschaften), funded by the German Federal Ministry of Education and Research (BMBF) and the German State Ministries for Research of Baden-Württemberg (MWK), Bayern (StMWFK), and Nordrhein-Westfalen (MIWF).

## References

1. Ewert, R., Schröder, W.: Acoustic perturbation equations based on flow decomposition via source filtering. J. Comput. Phys. **188**(2), 365–398 (2003). http://dx.doi.org/10.1016/S0021-9991(03)00168-2, https://doi.org/10.1016/S0021-9991(03)00168-2
2. Bridges, J., Brown, C.: Parametric testing of chevrons on single flow hot jets. In: AIAA Paper 2004–2824. http://dx.doi.org/10.2514/6.2004-2824, https://doi.org/10.2514/6.2004-2824
3. Niemöller, A., Meinke, M., Schröder, W., Albring, T., Gauger, N.: Noise reduction using a Direct-Hybrid CFD/CAA method, Paper 2019-2579. In: AIAA (2019). http://dx.doi.org/10.2514/6.2019-2579, https://doi.org/10.2514/6.2019-2579
4. Schlottke-Lakemper, M., Yu, H., Berger, S., Meinke, M., Schröder, W.: A fully coupled hybrid computational aeroacoustics method on hierarchical Cartesian meshes. Comput. Fluids **144**, 137–153 (2017). http://dx.doi.org/10.1016/j.compfluid.2016.12.001, https://doi.org/10.1016/j.compfluid.2016.12.001
5. Bailly, C., Juve, D.: Numerical solution of acoustic propagation problems using linearized Euler equations. AIAA J. **38**(1), 22–29 (2000). http://dx.doi.org/10.2514/2.949, https://doi.org/10.2514/2.949
6. Pitsch, H.: A consistent level set formulation for large-eddy simulation of premixed turbulent combustion. Combust. Flame **143**(4), 587–598 (2005)
7. Schlimpert, S., Feldhusen, A., Grimmen, J.H., Roidl, B., Meinke, M., Schröder, W.: Hydrodynamic instability and shear layer effects in turbulent premixed combustion. Phys. Fluids **28**(1), 017104 (2016)
8. Herff, S., Pausch, K., Schlimpert, S., Nawroth, H., Paschereit, C., Schröder, W.: Impact of burner plenum acoustics on the sound emission of a turbulent lean premixed open flame. Int. J. Spray Combust. (2020) (accepted for publication)
9. Pausch, K., Herff, S., Schröder, W.: Noise sources of an unconfined and a confined swirl burner. J. Sound. Vib. **475**(9), 115293 (2020)
10. Hartmann, D., Meinke, M., Schröder, W.: A strictly conservative Cartesian cut-cell method for compressible viscous flows on adaptive grids. Comp. Meth. Appl. Mech. Eng. **200**(9–12), 1038–1052 (2011). http://dx.doi.org/10.1016/j.cma.2010.05.015, https://doi.org/10.1016/j.cma.2010.05.015
11. Meinke, M., Schröder, W., Krause, E., Rister, T.: A comparison of second and sixth-order methods for large-eddy simulations. Comput. Fluids **31**(4–7), 695–718 (2002). http://dx.doi.org/10.1016/S0045-7930(01)00073-1, https://doi.org/10.1016/S0045-7930(01)00073-1
12. Boris, J.P., Grinstein, F.F., Oran, E.S., Kolbe, R.L.: New insights into large eddy simulation. Fluid Dyn. Res. **10**(4–6), 199–228 (1992). http://dx.doi.org/10.1016/0169-5983(92)90023-P, https://doi.org/10.1016/0169-5983(92)90023-P
13. Schneiders, L., Günther, C., Meinke, M., Schröder, W.: An efficient conservative cut-cell method for rigid bodies interacting with viscous compressible flows. J. Comput. Phys. **311**, 62–86 (2016). http://dx.doi.org/10.1016/j.jcp.2016.01.026, https://doi.org/10.1016/j.jcp.2016.01.026
14. Hartmann, D., Meinke, M., Schröder, W.: An adaptive multilevel multigrid formulation for Cartesian hierarchical grid methods. Comput. Fluids **37**(9), 1103–1125 (2008). http://dx.doi.org/10.1016/j.compfluid.2007.06.007, https://doi.org/10.1016/j.compfluid.2007.06.007
15. Günther, C., Meinke, M., Schröder, W.: A flexible level-set approach for tracking multiple interacting interfaces in embedded boundary methods, Comput. Fluids **102**, 182–202 (2014). http://dx.doi.org/10.1016/j.compfluid.2014.06.023, https://doi.org/10.1016/j.compfluid.2014.06.023

16. Kopriva, D., Woodruff, S., Hussaini, M.: Discontinuous spectral element approximation of Maxwell's equations. In: Discontinuous Galerkin Methods, Vol. 11 of LNCSE. Springer (2000). http://dx.doi.org/10.1007/978-3-642-59721-3_33, https://doi.org/10.1007/978-3-642-59721-3_33
17. Kopriva, D.A., Woodruff, S.L., Hussaini, M.: Computation of electromagnetic scattering with a non-conforming discontinuous spectral element method. Int. J. Numer. Meth. Eng. **53**(1), 105–222 (2002). http://dx.doi.org/10.1002/nme.394, https://doi.org/10.1002/nme.394
18. Carpenter, M.H., Kennedy, C.: Fourth-order 2N-storage Runge-Kutta schemes. NASA Report TM 109112, NASA Langley Research Center (1994)
19. Nourgaliev, R., Theofanous, T.: High-fidelity interface tracking in compressible flows: Unlimited anchored adaptive level set. J. Comput. Phys. **224**, 836–866 (2007)
20. Shu, C.-W., Osher, S.: efficient implementation of essentially non-oscillatory shock capturing schemes. J. Comput. Phys. **77**, 439–471 (1988)
21. Hartmann, D., Meinke, M., Schröder, W.: The constrained reinitialization equation for level set methods. J. Comput. Phys. **229**, 1514–1535 (2010)
22. Lintermann, A., Schlimpert, S., Grimmen, J.H., Günther, C., Meinke, M., Schröder, W.: Massively parallel grid generation on HPC systems. Comp. Meth. Appl. Mech. Eng. **277**, 131–153 (2014). http://dx.doi.org/10.1016/j.cma.2014.04.009, https://doi.org/10.1016/j.cma.2014.04.009
23. Pinar, A., Aykanat, C.: Fast optimal load balancing algorithms for 1D partitioning, J. Parallel Distrib. Comput. **64**(8), 974–996 (2004). http://dx.doi.org/10.1016/j.jpdc.2004.05.003, https://doi.org/10.1016/j.jpdc.2004.05.003
24. Miguet, S., Pierson, J.M.: Heuristics for 1D rectilinear partitioning as a low cost and high quality answer to dynamic load balancing. In: High-Performance Computing and Networking, Vol. 1225 of LNCS, pp. 550–564. Springer (1997). http://dx.doi.org/10.1007/BFb0031628, https://doi.org/10.1007/BFb0031628
25. Schlottke-Lakemper, , M., Niemöller, A., Meinke, M., Schröder, W.: Efficient parallelization for volume-coupled multiphysics simulations. Comp. Meth. Appl. Mech. Eng. **352** 461–487 (2019). http://dx.doi.org/10.1016/j.cma.2019.04.032, https://doi.org/10.1016/j.cma.2019.04.032
26. Hendrickson, B., Devine, K.: Dynamic load balancing in computational mechanics. Comp. Meth. Appl. Mech. Eng. **184**(2–4), 485–500 (2000). http://dx.doi.org/10.1016/S0045-7825(99)00241-8, https://doi.org/10.1016/S0045-7825(99)00241-8
27. DeRose, L., Homer, B., Johnson, D.: Detecting application load imbalance on high end massively parallel systems. In: Parallel Processing, vol. 4641 of LNCS, pp. 150–159. Springer (2007). http://dx.doi.org/10.1007/978-3-540-74466-5-17, https://doi.org/10.1007/978-3-540-74466-5-17
28. Böhme, D.: Characterizing Load and Communication Imbalance in Parallel Applications, vol. 23 of IAS, Forschungszentrum Jülich (2014). http://dx.doi.org/10.1109/IPDPSW.2012.321, https://doi.org/10.1109/IPDPSW.2012.321
29. Teresco, J.D., Devine, K.D., Flaherty, J.E.: Partitioning and dynamic load balancing for the numerical solution of partial differential equations. In: Numerical Solution of Partial Differential Equations on Parallel Computers, vol. 51 of LNCSE, pp. 55–88. Springer (2006). http://dx.doi.org/10.1007/3-540-31619-1-2, https://doi.org/10.1007/3-540-31619-1-2
30. Hendrickson, B.: Load balancing fictions, falsehoods and fallacies. Appl. Math. Model. **25**(2), 99–108 (2000). http://dx.doi.org/10.1016/S0307-904X(00)00042-1, https://doi.org/10.1016/S0307-904X(00)00042-1
31. Watts, J., Taylor, S.: A practical approach to dynamic load balancing. IEEE Trans. Parallel Distrib. Syst. **9**(3), 235–248 (1998). http://dx.doi.org/10.1109/71.674316, https://doi.org/10.1109/71.674316
32. Jetley, P., Gioachin, F., Mendes, C., Kalé, L., Quinn, T.: Massively parallel cosmological simulations with ChaNGa. In: IPDPS, pp. 1–12. IEEE (2008). http://dx.doi.org/10.1109/IPDPS.2008.4536319, https://doi.org/10.1109/IPDPS.2008.4536319

33. Dongarra, J., et al.: The international exascale software project roadmap. Int. J. High Perform. Comput. Appl. **25**(1), 3–60 (2011). http://dx.doi.org/10.1177/1094342010391989, https://doi.org/10.1177/1094342010391989
34. Menon, H., Jain, N., Zheng, G., Kalé, L.: Automated load balancing invocation based on application characteristics. in: Cluster Comput. 373–381. IEEE (2012). http://dx.doi.org/10.1109/CLUSTER.2012.61, https://doi.org/10.1109/CLUSTER.2012.61
35. Anderson, E., Bai, Z., Bischof, C., Blackford, S., Demmel, J., Dongarra, J., Du Croz, J., Greenbaum, A., Hammarling, S., McKenney, A., Sorensen, D.: LAPACK Users' Guide, 3rd edn. SIAM (1999)
36. Tanna, H.K.: An experimental study of jet noise Part I: Turbulent mixing noise. J. Sound Vibr. **50**(3), 405–428 (1977). http://dx.doi.org/10.1016/0022-460X(77)90493-X, https://doi.org/10.1016/0022-460X(77)90493-X
37. Pauz, V., Niemöller, A., Meinke, M., Schröder, W.: Numerical analysis of chevron nozzle noise. In: AIAA Paper 2017–3853. http://dx.doi.org/10.2514/6.2017-3853, https://doi.org/10.2514/6.2017-3853
38. Moeck, J.P., Bourgouin, J.-F., Durox, D., Schuller, T., Candel, S.: Nonlinear interaction between a processing vortex core and acoustic oscillations in a turbulent swirling flame. Combust. Flame **159**(8), 2650–2668 (2012)
39. Gupta, A., Lilley, D., Syred, N.: Swirl Flows. Abacus Press, Tunbridge Wells, England (1984)
40. Poinsot, T., Lele, S.K.: Boundary conditions for direct simulations of compressible viscous flows. J. Comput. Phys. **101**(1), 104–129 (1992)
41. Rudy, D.H., Strikwerda, J.C.: A nonreflecting outflow boundary condition for subsonic Navier-stokes calculations. J. Comput. Phys. **36**, 55–70 (1980)

# Brinkman Penalization and Boundary Layer in High-Order Discontinuous Galerkin

**Neda Ebrahimi Pour, Nikhil Anand, Felix Bernhards, Harald Klimach, and Sabine Roller**

**Abstract** In this chapter we look into a high-order representation of complex geometries by the Brinkman penalization method. We focus on the effect of this immersed boundary model on the boundary layers in a high-order discontinuous Galerkin scheme. High-order approximations are attractive on modern computing architectures, as they require few degrees of freedom, to represent smooth solutions, resulting in a smaller memory footprint when compared to lower order discretizations. A significant hurdle in using high-order methods for problems involving complex geometries is matching the surface description of the geometry with the discretization scheme. The Brinkman penalization offers a method to achieve this without the need for complicated mesh generation and special elements. We investigated its use in our high-order discontinuous Galerkin implementation Ateles in [2], where we looked at inviscid effects like reflections. Here we investigate the viscous boundary layer close to a wall, modeled by the penalization.

## 1 Introduction

Numerical simulations have been established as a central pillar in engineering applications, and modern supercomputing facilities allow for the inclusion of more and more physical aspects in these simulations. Engineering problems typically involve complex geometries that need to be represented accordingly in the numerical model. Intensive studies have been conducted to model complex geometries with high accu-

N. Ebrahimi Pour (✉) · N. Anand · F. Bernhards · H. Klimach · S. Roller
University of Siegen, Adolf-Reichwein-Str. 2, 57072 Siegen, Germany
e-mail: neda.epour@uni-siegen.de

N. Anand
e-mail: nikhil.anand@uni-siegen.de

H. Klimach
e-mail: harlad.klimach@uni-siegen.de

S. Roller
e-mail: sabine.roller@uni-siegen.de



M. M. Resch et al. (eds.), *Sustained Simulation Performance 2019 and 2020*,
https://doi.org/10.1007/978-3-030-68049-7_6

racy, while maintaining efficient computation. An overview can be found e.g., in [3], where fluid-structure interactions are discussed. We are here concerned with flow simulations, as these still pose a significant computational challenge. One can categorize the efforts to represent complex geometries into body conforming and non-conforming mesh methods. In the first case, the geometry is integrated into the computational mesh and the mesh is aligned to the geometry. Hence, very small elements can be found close to the geometry interface, while away from the interface the computational mesh elements become larger in size. This strategy allows to resolve the boundary layer at the geometrical interface. The second method is also well known as immersed boundary method. Here, mostly a Cartesian mesh is used for the numerical computation, while the geometry is embedded into the governing equations. This technique allows to model the geometry with the same numerical discretization method as the flow field. The non-conforming mesh methods have a major benefit for moving geometries, where the mesh remains unchanged over the entire simulation time, while in the case of the conforming mesh method, the mesh needs to be re-generated and updated with the movement of the geometry. Some extended discussion on those two methods can be found for example in [6].

In this work we investigate non-moving geometries, which are modelled as an artificial porous material, by the method known as the Brinkman penalization. Literature on this method is, albeit using the Navier-Stokes equations, mostly concerned with aeroacoustic problems such as found in Liu et al. [11] or Komatsu et al. [9]. However, investigations related to the boundary layer have, to our knowledge, not been conducted so far. This investigation is of importance, as boundary layers might have a great impact on numerical simulations involving e.g., the wing stall or the emission of noise. Therefore, in this work, we present first results on the boundary layer investigation, when modelling geometries as an artificial porous material. The structure of this work is as follows: We first introduce the modelling method used to model arbitrary geometrical shapes. Afterward, we present first results of a laminar channel flow and compare it against a reference solution, obtained through imposing a traditional wall boundary condition. Additionally, we also investigate a curved boundary test case, to demonstrate that this method indeed is suitable to represent such geometries accurately in the high-order scheme. Lastly, we conclude and summarize our outcomes.

## 2 Methodology

In this section we briefly recall the Brinkman penalization technique, used to model arbitrary geometrical shapes. The volume penalization method introduces additional artificial terms to the governing equations to be solved in areas where geometries are present. Those terms appear as source terms on the right hand sides of the conservation equations. The compressible Navier-Stokes equations with the penalization terms are presented in (1), see also [9].

$$\partial_t \rho + \nabla \cdot \boldsymbol{m} = -\boxed{\left(\frac{1}{\phi} - 1\right) \chi \frac{\partial (u_i - U_{oi})\rho}{\partial x_i}}, \tag{1a}$$

$$\partial_t m_i = -\frac{\partial}{\partial x_j}\left(m_i u_j\right) - \frac{\partial p}{\partial x_i} + \frac{\partial \tau_{ij}}{\partial x_j} - \boxed{\frac{\chi}{\eta}\left(u_i - U_{oi}\right)}$$
$$- \boxed{u_i \left(\frac{1}{\phi} - 1\right) \chi \frac{\partial}{\partial x_j} [\rho (u_i - U_{oi})]}, \tag{1b}$$

$$\partial_t E = -\frac{\partial}{\partial x_j}\left[(E + p)u_j\right] + \frac{\partial}{\partial x_i}(u_i \tau_{ij}) + \frac{\partial}{\partial x_j}\left(k \frac{\partial T}{\partial x_j}\right) - \boxed{\frac{\chi}{\eta_T}(T - T_o)}$$
$$- \boxed{\frac{\chi}{\eta}(\boldsymbol{u} - \boldsymbol{U}_o) \cdot \boldsymbol{u} - \frac{|\boldsymbol{u}|^2}{2}\left(\frac{1}{\phi} - 1\right) \chi \frac{\partial}{\partial x_j}[\rho (u_i - U_{oi})]}, \tag{1c}$$

Here $\eta$ and $\eta_T$ represent the viscous and thermal permeability respectively and $\phi$ the porosity of the Brinkman model. The velocity and temperature of the obstacle are given by $U_o$ and $T_o$, and its location is defined by the masking function $\chi$. While the permeability terms only appear in the source terms, the porosity term affects the divergence operators and, thus, changes the eigenvalues of the convective system. This results in smaller time steps in the explicit time integration scheme we use, due to its stability limit.

The porosity $\phi$, introduced by Liu and Vasilyev [11] for compressible flows, however, can be set to $\phi = 1$, if the permeabilities are chosen sufficiently small, as shown in [2]. Tiny permeabilities turn the problem stiff, but as these appear as purely local source terms, we can employ an implicit time-integration for just those terms with an implicit mixed explicit time integration scheme. The overall computation remains explicit and the implicit part can be efficiently computed locally.

Neglecting the porosity by setting $\phi = 1$, the equations in (1) are simplified to:

$$\partial_t \rho = -\frac{\partial m_i}{\partial x_i} \tag{2a}$$

$$\partial_t m_i = -\frac{\partial}{\partial x_j}\left(m_i u_j\right) - \frac{\partial p}{\partial x_i} + \frac{\partial \tau_{ij}}{\partial x_j} - \boxed{\frac{\chi}{\eta}\left(u_i - U_{oi}\right)} \tag{2b}$$

$$\partial_t E = -\frac{\partial}{\partial x_j}\left[(E + p)u_j\right] + \frac{\partial}{\partial x_i}(u_i \tau_{ij}) + \frac{\partial}{\partial x_j}\left(k \frac{\partial T}{\partial x_j}\right) - \boxed{\frac{\chi}{\eta_T}(T - T_o)}$$
$$\boxed{-\frac{\chi}{\eta}(\boldsymbol{u} - \boldsymbol{U}_o) \cdot \boldsymbol{u}}. \tag{2c}$$

We deploy the diagonally implicit Runge-Kutta scheme [1] to solve the source terms implicitly. The following permeability terms, proposed by Anand et al. [2], are used to model obstacles. We define the viscous permeability $\eta = \beta^2$ and the thermal viscosity $\eta_T = 0.4\beta$. The expected modelling error has than, according to the analysis in [11] a magnitude of $\beta^{1/4}$ for a porosity of 1.0. For our analysis here, we use a scaling factor of $\beta \leq 10^{-6}$. The conservation equations are discretized with a high-order discontinuous Galerkin method [5].

## 3 Results

We now turn to the investigation of the influence of the described wall model on the boundary layer at the wall in the viscous flow. For this investigation we first use a simple channel flow. As, the Hagen-Poiseuille flow is only valid in the inviscid case, and most references in literature refer to incompressible solutions, we employ a computation with a classical wall boundary condition as a reference in our comparisions. Thereafter, we investigate a curved boundary test case of flow past a circular cylinder and compare the results to previously published results.

### 3.1 Flow Through a Channel

**Test case description**: The two-dimensional channel is set up with a height of 1 m and a length of 2 m. The boundary at the top is a no-slip wall and the bottom wall is modelled using the Brinkman penalization technique to resemble a solid wall. The reference solution is obtained by setting up a channel using traditional wall boundary conditions for top and bottom walls. The bottom wall in the reference solution is chosen to be isothermal, since the temperature of the wall, modelled by the Brinkman penalization, is constant. We perform simulations with different scheme orders and mesh resolutions, to investigate how the error in comparison to the reference solution behaves. The density $\rho$ is defined as 1 kg/m$^3$ and the pressure $p$ is equal to 1 bar. The Mach number is 0.3 and a Reynolds number of 100 is chosen with respect to the height of the channel. The low Reynolds number should ensure a laminar flow. The time step size is controlled by the Courant-Friedrichs-Lewy (CFL) condition and we use a Courant factor of 0.3. A vertical line, located at $x = 1.0$, along the height of the channel is tracked. The simulation runs until a steady state is attained.

*Initial condition*: Initially a pressure drop according to [10] is prescribed,

$$p_A = 12 \cdot u_x \frac{\eta}{h^2} \cdot \frac{l - x}{l}, \tag{3}$$

with the velocity $u_x$ in x-direction, that can be computed from the Mach number, the kinematic viscosity $\eta$ and $h$ and $l$ are the height and length of the channel, respectively. We also initially prescribe the Hagen-Poiseuille velocity profile in x-direction, that is defined as

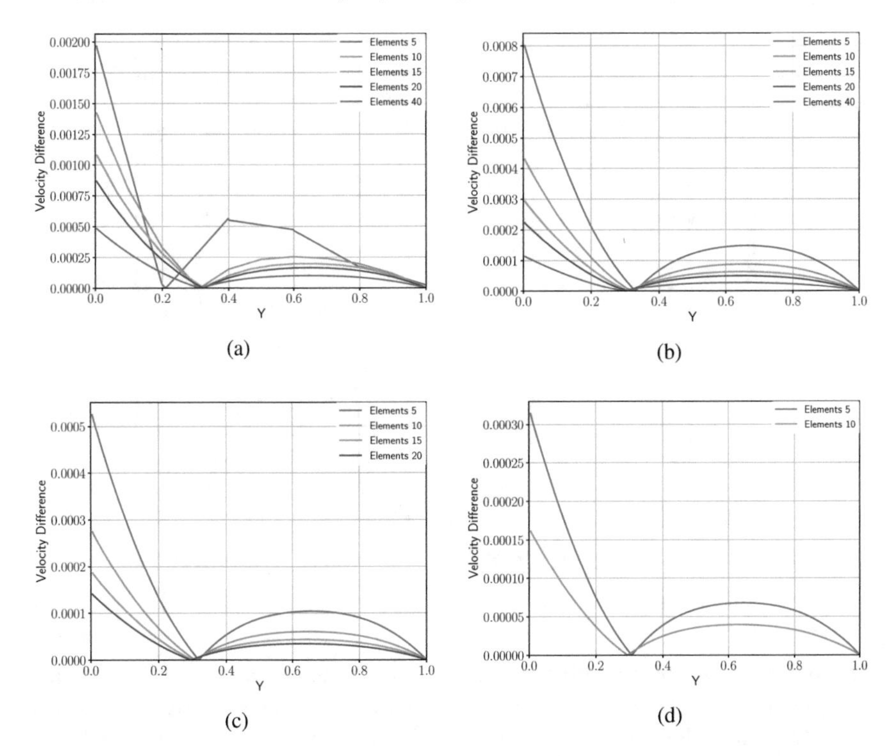


**Fig. 1** Comparison of the Velocity: Difference between the reference solution (isothermal boundary condition) and the modelled porous wall, for $h$ refinement investigation. The investigated scheme orders are (**a**) $O(3)$, (**b**) $O(7)$, (**c**) $O(9)$ and (**d**) $O(12)$. With the no-slip wall located at $y = 1.0$ and the modelled porous wall at $y = 0.0$

$$u_A = \frac{y \cdot (h - y)}{2\eta} \cdot \frac{dp}{dx}. \tag{4}$$

where $\frac{dp}{dx}$ is the pressure gradient in x-direction. The velocity in y-direction is set to 0.

*Boundary Conditions*: For the left and the right boundary, we prescribe Dirichlet boundary conditions, where the pressure and velocity profiles from the initial conditions are defined at the boundaries. For the top boundary a no-slip wall is used and at the bottom boundary, as stated, a wall modelled by the penalization technique is prescribed.

Figure 1 shows the error in velocity along the height of the channel, for different scheme orders. Different curves in the figure represent the refinement of the mesh ($h$ refinement) along the channel height. Note, that the position of the modelled wall is at $y = 0$. From Figure 1a, we can observe, that with higher mesh resolution, the error in velocity close to the wall, i.e., at $y = 0$ becomes smaller. With the highest element count along the height of the channel, i.e., 40 elements, the error stands at

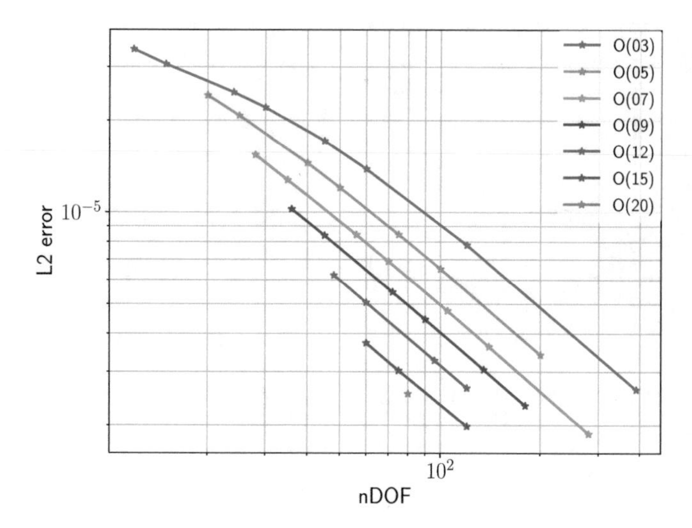


**Fig. 2** L2 error plot over the degrees of freedom (nDof) for different scheme orders $O$ and element counts

around 0.0005, which is significantly close to the reference solution. Furthermore, with increasing orders, we observe in figures Fig. 1b, c and d, we need a fewer number of elements in the height of the channel to achieve errors which are even smaller. In the case of $O(7)$ only $7 - 8$ elements in the height would be sufficient, and in the case of $O(9)$ and $O(12)$ roughly $4 - 5$ and $2 - 3$ elements respectively would be adequate to reach the same error level. Viewing this in terms of number of degrees of freedom (nDof), high-order methods would attain the same accuracy using less degrees of freedom. For instance in case of $O(3)$ and 40 elements in the height (cf. Figure 1a), this would lead to 80 elements lengthwise (channel being twice as long compared to its height) and result in 28800 degrees of freedom for each variable. In contrast, computing the same problem with a scheme order of $O(12)$ (cf. Figure 1b), only 3 elements in the height are sufficient to achieve the same error, and this would only be 2592 degrees of freedom per variable. Therefore, it is obvious that with higher scheme orders, the memory required for a given accuracy is significantly reduced. For example, in the case described above, using the $12^{th}$ order $O(12)$ would lead to a 90% reduction in memory consumption.

Figure 2 shows this trend clearly when we compare the L2 error of the modelled wall over the number of degrees of freedom. Each line represents the $h$ refinement for a different scheme order with number of degrees of freedom on the x-axis.

In Figure 3 the velocity profile along the y-axis is shown for the case $O(3)$ and $O(12)$ with the same number of degrees of freedom of 7200 per variable. Figure 3b shows the zoomed in view of Figure 3a, here the error in the velocity profile is clearly visible. Even though, both solutions were computed using the same number of degrees of freedom, the solution obtained using high-order, namely $O(12)$, is closer to the desired reference solution than $O(3)$.

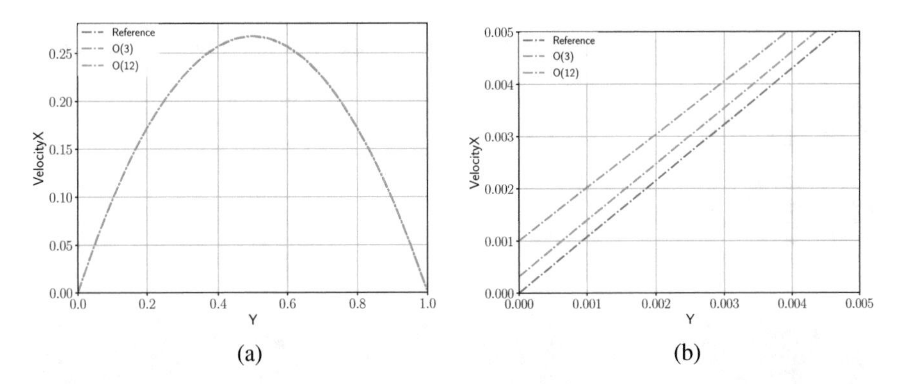


**Fig. 3** Comparison: Velocity profile along the height of the channel, when compared to the reference solution for $O(3)$ and $O(12)$ for 7200 Dofs. The velocity profile is shown in **(a)** and in **(b)** a zoom-in of the location $y = 0.0$, where the wall is modelled as a porous material

From our investigation in this section, we conclude that the modelled wall is in very good agreement with the reference solution, obtained using traditional wall boundary conditions. This further establishes that the boundary layer close to the wall is comparable and in agreement with traditional wall boundary conditions. Further, we demonstrated, less number of degrees of freedom are needed when using higher orders to attain a certain accuracy. This has an advantage of less memory consumption and is favourable when considering large-scale simulations on modern supercomputers, where memory bandwidth and capacity are limiting factors. In the next section, we investigate the flow over a cylinder, and compare our solution with those found in literature.

## *3.2 Flow Past a Circular Cylinder*

After demonstration of basic numerical and convergence properties of the penalization scheme with the channel flow, we now investigate a two dimensional flow past a circular cylinder. In general, curved mesh representation gets problematic for high-order schemes. Also, performing this investigation would further facilitate the identification of numerical issues introduced by curved walls. To validate our results, we use the extensive numerical and experimental datasets which already exists for this kind of flow e.g., in [4, 8, 13].

**Test case description**: The Reynolds number of the inflow is chosen as $Re = 100$ and is defined as

$$Re = \frac{u\rho d}{\nu}, \tag{5}$$

where, $u$ is the free-stream velocity, $\rho$ is the background density, $d$ denotes the diameter of the cylinder and $\nu$ is the viscosity. The simulation is performed in a

**Table 1** Comparison of the drag coefficient, $C_d$ and Strouhal number, $St$ for $Re = 100$ against established results

| Contributions | $C_d$ | $St$ |
|---|---|---|
| Tseng Ferziger [13] | 1.42 | 0.164 |
| Kim et al. [8] | 1.33 | 0.165 |
| De Palma et al. [4] | 1.32 | 0.163 |
| Present work | 1.40 | 0.160 |

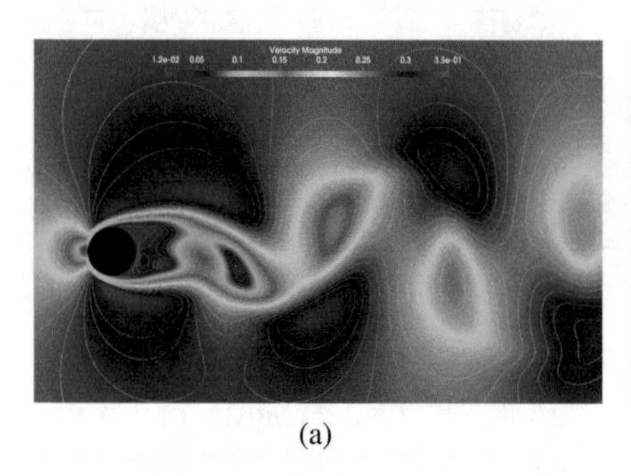


(a)

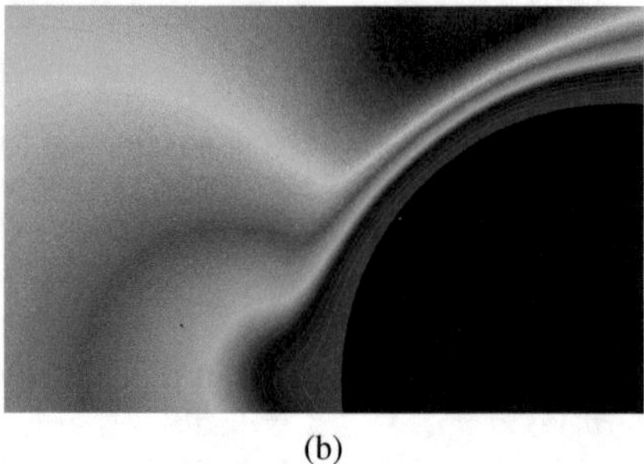

(b)

**Fig. 4** View of the velocity profile and contours around the cylindrical obstacle captured at an instant. Figure (**b**) shows further closeup view of the area around the cylinder. In this zoomed views, also attached boundary layers, separated shear layers as well as the formation and evolution of the Karman vortices in the wake of the cylinder at $Re = 100$ can be observed

sufficiently large domain of size, $\Omega = 28d \times 40d$, to minimize the outer boundary effects. Cubical elements with an edge length of $dx = d/8$ are used to discretize the complete domain. Results are obtained solving the compressible Navier-Stokes equations in two dimensions using a spatial scheme order of $O(8)$, i.e., each variable is represented by 64 degrees of freedom per element.

According to the literature, the re-circulation region around the wake of a cylinder, immersed in a free-stream, becomes unstable and starts oscillating around the critical Reynolds number of $Re = 46$ [7, 12]. This is also what we expect to observe at a higher Reynolds number chosen for this simulation, i.e., $Re = 100$. Once the flow attains a stable state, we compute the drag coefficient, $C_d$, defined as $C_d = \frac{F_d}{0.5\rho u^2 d}$, where $F_d$ is the drag force. The results obtained in terms of dimensionless quantities, i.e., mean drag coefficient, $C_d$ and Strouhal number, $St$ are summarized and compared with the reference results in Table 1. We see that the results closely predict both the flow properties and shows an excellent agreement with the previously obtained experimental and numerical results. Figure 4 shows the contour plot of the velocity magnitude around the cylinder captured at an instant of time. At this resolution, we can observe smooth attached boundary layers as well as separated shear layers. Also, in the wake of the cylinder, the formation and evolution of Karman vortices can be recognized.

## 4 Conclusion

An investigation of the shear layers in compressible flow simulations with complex geometries using the Brinkman penalization method is presented. For validation we use walls of a channel and compare against the reference solution and show the behaviour of error with different scheme orders. We also demonstrated, that the penalization method used in this work, provides small errors and convergences towards the reference solution much faster, with higher polynomial degrees and less number of degrees of freedom. This has the major advantage, that less degrees of freedom are necessary to achieve a small error, resulting in less memory consumption for the simulation. The real advantage of the method, however, is representing curved boundaries in multiple dimensions. We presented a uniform flow past a stationary cylinder at $R_e = 100$. The numerical solution showed an excellent agreement when compared with published results. With the presented method, it is therefore possible to represent complex geometries, that are consistent with the high-order scheme. Furthermore, an extension of this method to problems involving moving geometries is currently being investigated.

**Acknowledgements** Neda Ebrahimi Pour was financially supported by the priority program 1648—Software for Exascale Computing 214 (www.sppexa.de) of the German Research Foundation.

## References

1. Alexander, R.: Diagonally implicit rungekutta methods for stiff O.D.E.s. SIAM J. Numer. Anal. **14**(6), 1006–1021 (1977). https://doi.org/10.1137/0714068
2. Anand, N., Ebrahimi Pour, N., Klimach, H., Roller, S.: Utilization of the brinkman penalization to represent geometries in a high-order discontinuous galerkin scheme on octree meshes. Symmetry **11**(9), 1126 (2019). https://doi.org/10.3390/sym11091126
3. Bungartz, H.J., Mehl, M., Michael, S.: Fluid Structure Interaction II. Springer, Berlin (2010)
4. De Palma, P., de Tullio, M., Pascazio, G., Napolitano, M.: An immersed-boundary method for compressible viscous flows. Comput. Fluids **35**(7), 693–702 (2006). https://doi.org/10.1016/j.compfluid.2006.01.004, http://www.sciencedirect.com/science/article/pii/S0045793006000065. Special Issue Dedicated to Professor Stanley G. Rubin on the Occasion of his 65th Birthday
5. Hesthaven, J.S., Warburton, T.: Nodal Discontinuous Galerkin Methods: Algorithms, Analysis, and Applications, 1st edn. Springer Publishing Company, Incorporated (2007)
6. Hou, G., Wang, J., Layton, A.: Numerical methods for fluid-structure interaction a review. Commun. Comput. Phys. **12**(2), 337–377 (2012). https://doi.org/10.4208/cicp.291210.290411s
7. Jackson, C.P.: A finite-element study of the onset of vortex shedding in flow past variously shaped bodies. J. Fluid Mech. **182**, 23–45 (1987). https://doi.org/10.1017/S0022112087002234
8. Kim, J., Kim, D., Choi, H.: An immersed-boundary finite-volume method for simulations of flow in complex geometries. J. Comput. Phys. **171**(1), 132–150 (2001). https://doi.org/10.1006/jcph.2001.6778, http://www.sciencedirect.com/science/article/pii/S0021999101967786
9. Komatsu, R., Iwakami, W., Hattori, Y.: Direct numerical simulation of aeroacoustic sound by volume penalization method. Comput. Fluids **130**, 24–36 (2016). https://doi.org/10.1016/j.compfluid.2016.02.016

10. Landau, L.D., Lifshitz, E.M.: Fluid Mechanics. Elsevier Butterworth-Heinemann (2012)
11. Liu, Q., Vasilyev, O.V.: A brinkman penalization method for compressible flows in complex geometries. J. Comput. Phys. **227**(2), 946–966 (2007). https://doi.org/10.1016/j.jcp.2007.07.037
12. Provansal, M., Mathis, C., Boyer, L.: Bénard-von kármán instability: transient and forced regimes. J. Fluid Mech. **182**, 1–22 (1987). https://doi.org/10.1017/S0022112087002222
13. Tseng, Y.H., Ferziger, J.H.: A ghost-cell immersed boundary method for flow in complex geometry. J. Comput. Phys. **192**(2), 593–623 (2003). https://doi.org/10.1016/j.jcp.2003.07.024, http://www.sciencedirect.com/science/article/pii/S0021999103004108

# Data Handling and New Concepts

# Handling Large Numerical Data-Sets: Viability of a Lossy Compressor for CFD-simulations

**Patrick Vogler and Ulrich Rist**

**Abstract** Over the years, a steady increase in computing power has enabled scientists and engineers to develop increasingly complex applications for machine learning and scientific computing. But while these applications promise to solve some of the most difficult problems we face today, their data hunger also reveals an ever-increasing I/O bottleneck. It is therefore imperative that we develop I/O strategies to better utilize the raw power of our high-performance machines and improve the usability and efficiency of our tools. To achieve this goal, we have developed the BigWhoop compression library based on the JPEG 2000 standard. It enables the efficient and lossy compression of numerical data-sets while minimizing information loss and the introduction of compression artifacts. This paper presents a comparative study using the Taylor-Green Vortex test case to demonstrate the superior compression performance of BigWhoop compared to contemporary solutions. An evaluation of compression-related distortion at high compression ratios is shown to prove its feasibility for both visualization and statistical analysis.

## 1 Introduction

One of the hallmarks of High Performance Computing (HPC) is its ability to perform complex numerical simulations. The compute-intensive workload is divided into small, manageable junk that can be distributed to highly parallel computer clusters. This approach, combined with a steady increase in single-core performance, has enabled the scientific community to study increasingly complex physical problems on increasingly more powerful systems. However, it is becoming more and more clear that the path to exascale performance is blocked by an I/0 bottleneck. Modern high-performance systems are capable of generating and processing large amounts of data

P. Vogler (✉)
High Performance Computing Center, Nobelstraße 19, 70569 Stuttgart, Germany
e-mail: hpcpvogl@hlrs.de

U. Rist
Institut of Aerodynamics and Gasdynamics, Pfaffenwaldring 21, 70569 Stuttgart, Germany
e-mail: ulrich.rist@iag.uni-stuttgart.de



M. M. Resch et al. (eds.), *Sustained Simulation Performance 2019 and 2020*,
https://doi.org/10.1007/978-3-030-68049-7_7

quickly. Overall performance, on the other hand, is often hampered by how quickly the system can transfer and store calculated data. Given that researchers are constantly trying to study simulations with ever higher spatial and temporal resolution, the increase in computing power will therefore only exacerbate this problem [5, 8]. Our best way forward should therefore be to use the otherwise wasted computing cycles to reduce the file size of our numerical simulations.

Since effective data storage is an ubiquitous IT problem, much effort has already been put into the development and refinement of compression algorithms. However, most of the common compression techniques available for floating point numbers are so-called lossless dictionary encoders (i.e., the Lempel-Ziv-Welch algorithm [14]). Since they are not capable of neglecting parts of the original data that contribute little to the overall information content, they are limited to a size reduction of 10–30% [11]. Furthermore, these encoders only affect the statistical redundancies of the underlying bitstream and are not able to exploit the spatially correlated information. However, considering that most physical problems are subject to diffusion, our numerical data-sets are usually smooth and continuous, resulting in a frequency spectrum dominated by lower modes [10]. Prominent algorithms that exploit these spatial redundancies and enable lossy compression can be found in the world of entertainment technology.

The goal of this work is the development of a library for lossy compression of structured numerical data-sets. The rate control solution needs to conserve most of the internal energy of the physical domain and minimize the introduction of compression artifacts to support both visual and statistical evaluation. To allow for easier storage, the compressed bit-stream is required to embed all the information needed to decompress the simulation file. To this end, the compression library described in this paper has been derived from the JPEG 2000 (JP2) standard.

A review of similar approaches to adapt image compression standards for floating-point values is presented in Sect. 2. A technical description of our compression library will follow in Sect. 3. An evaluation of the compression-induced distortion on a numerical simulation of the Taylor-Green vortex decay is presented in Sect. 4. Finally, we conclude with some closing remarks in Sect. 5.

## 2 Previous Work

In this section, we want to give an overview of some existing approaches for adapting image compression ideas to floating point data-sets. It is not our intention to present a complete treatise on all existing work on this topic, but rather to highlight the shortcomings of some of the better known ideas.

Loddoch and Schmalzl [8] have extended the Joint Photographic Experts Group (JPEG) standard to volumetric floating point arrays by applying the one-dimensional real to real discrete cosine transform (DCT) along the axis of each spatial dimension. The DCT coefficients are then quantized and encoded using a variable length code similar to that proposed by the JPEG standard.

Lindstrom [6], on the other hand, uses the fixed-point number-format Q, which maps the floating-point values onto the dynamic range of a specified integer type. A lifting based integer-to-integer block transform is then applied to data blocks comprising of 4 samples in every dimension. Next, the transform coefficients are encoded using an embedded coding algorithm to produce a quality-scalable code-stream. Both compression algorithms are simple and efficient in exploiting the low-frequency nature of most numerical data-sets. The non-local basis functions of the respective transforms, however, will result in large-scale compression artifacts which heavily distort the data-set [1].

The JPEG 2000 standard has been introduced to avert the introduction of these so called blocking artifacts. In contrast to the base-line JPEG standard, JPEG 2000 employs a lifting-based one-dimensional discrete wavelet transform (DWT) that can be performed by either the reversible LeGall-5/3-Wavelet (5/3-CDF-Wavelet) for lossless or the non-reversible Cohen-Daubechies-Feauveau-9/7-Wavelet (9/7-CDF-Wavelet) filter for lossy coding [11].

Efforts have already been made to adapt the full range of JP2 features to floating-point numbers. Usevitch [12] proposed an extended integer representation of single precision IEEE 754 floating-point numbers which stores the 24 consecutive mantissa bits losslessly using 278 bit locations. While this approach requires little alteration of the base code, the extended integer representation results in a significant increase in memory consumption and overall decrease in compression performance.

Gamito and Salles Dias [4], on the other hand, split the floating-point values into their sign, bit and mantissa fields and processed them separately. The so-called shape-adaptive discrete wavelet transform is applied inside the smooth regions of the mantissa field to avoid high-frequency information that is introduced during the splitting operation. The increase in computational complexity the shape-adaptive wavelet transform incurs, however, limits the usefulness of this approach for large-scale simulations. We therefore propose to adapt a more memory efficient approach, combining the JPEG 2000 standard with the fixed-point number-format Q. This idea was forged into the new compression library BigWhoop which we want to describe in the following.

## 3 Theory

The compression library described in this paper has been derived from the JPEG 2000 standard. The standard offers an embedded block coding algorithm that generates a quality- and resolution-scalable bit-stream which is optimally truncated with regards to the compression induced distortion. Furthermore, JPEG 2000 has been designed around the discrete wavelet transform, allowing for random access as well as region-dependent rate-distortion-control operations [11]. Combined, these features result in an algorithm that can efficiently compress a wide variety of numerical problems without sacrificing crucial information.

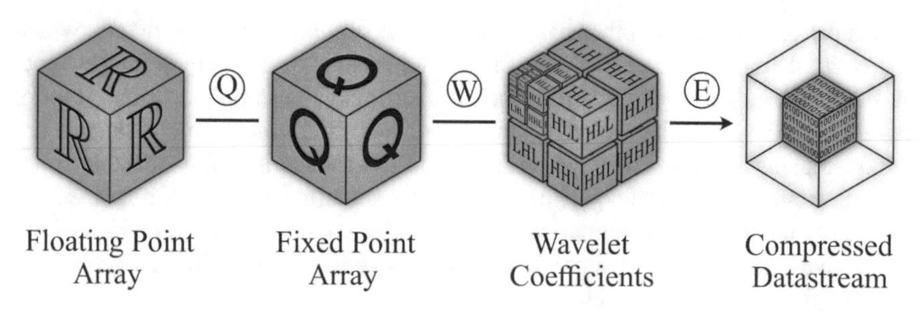


**Fig. 1** Structure of wavelet-based compression algorithm for volumetric floating-point arrays. Encircled letters indicate the floating-point to fixed-point transform (Q), discrete wavelet transform (W) and entropy encoding stage (E)

The fundamental structure of our compression library, dubbed BigWhoop, is shown in Fig. 1. Since it has been adapted from the JP2 standard, our first course of action was to transform the volumetric floating-point array into a format suitable for the compressor. This was accomplished by using the *fixed-point number format Q*. The next step was to generate a time-frequency representation of the transformed data samples. Here, the *discrete wavelet transform* is applied to the numerical field to decorrelate its inherent spatial frequency information. Finally, the wavelet coefficients are quantized and *entropy encoded* using the Embedded Block Coding with Optimized Truncation (EBCOT) algorithm. For a thorough discussion of the JPEG 2000 standard the reader is referred to the treatise by Taubman and Marcellin [11].

**Fixed-Point Number Format Q** In numerical analysis, floating-point numbers are adopted to approximate the large dynamic range of real numbers used to describe various physical phenomena. The non-linearity of the mantissa, however, will introduce high-frequency information into the binary representation of the numerical values. This, in turn, can degrade the performance of a compression algorithm [4, 12]. To bypass this problem, the number format Q is used to transform the data samples to a fixed-point representation with constant resolution [7, 15]. To this end, the dynamic range of each flow-field variable is first centered around zero. This is done to reduce the number of significant bits needed to represent the data samples and exploit the full range of the number format Q. Next, all values are normalized to the range $[-1, +1]$ and then multiplied by the number of fractional bits $Qm$ used in the fixed-point representation. In this context, the number of fractional bits should not exceed the width of the integer type used to store the fixed-point values. For our purpose, we use $Qm = 32$ to allow for a fast compression scheme.

**Discrete Wavelet Transform** After transforming the volumetric floating-point into a fixed-point array, the lifting-based, one-dimensional discrete wavelet transform is applied to our data-set. The DWT is responsible for transforming the fixed-point array into a time-frequency representation. This step will concentrate the internal information into a small number of wavelet coefficients and extract the location of high-frequency information that can be omitted during rate control.

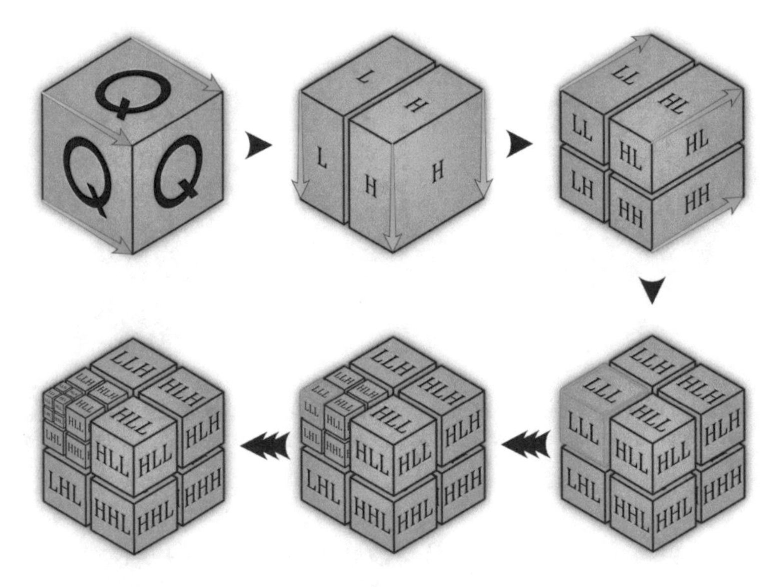


**Fig. 2** One-dimensional discrete wavelet transform applied to a 2-dimensional data-set. Blue lines indicate directions of one-dimensional transform. Letters H and L imply high-and low-frequency bands. Subscripts signal the resolution level

The forward wavelet transform is best understood as a pair of low- and high-pass filters, commonly known as the analysis filter-bank. The filter bank is followed by a down-sampling operation by a factor of two, which is necessary to ensure a critically sampled signal that is split into a low- and high-frequency band (see Fig. 2).

The *low-pass filter* attenuates high-frequency information, creating a blurred version of the original data-set. This low-frequency band should represent a highly correlated signal which can be subjected to further wavelet decompositions, producing a so-called dyadic decomposition.

The *high-pass filter*, on the other hand, preserves the high-frequency information that has been discarded in the low-frequency band. Preferably, the high-frequency band is sparsely occupied, resulting in a highly decorrelated signal [1, 3].

The discrete wavelet transform implemented in our compression library is the 9-tab/7-tab real-to-real filter bank, commonly known as the 9/7-CDF-Wavelet. The 9/7-filter bank is split into two predictor (high-band) and two update (low-band) operations, followed by a dual normalization step:

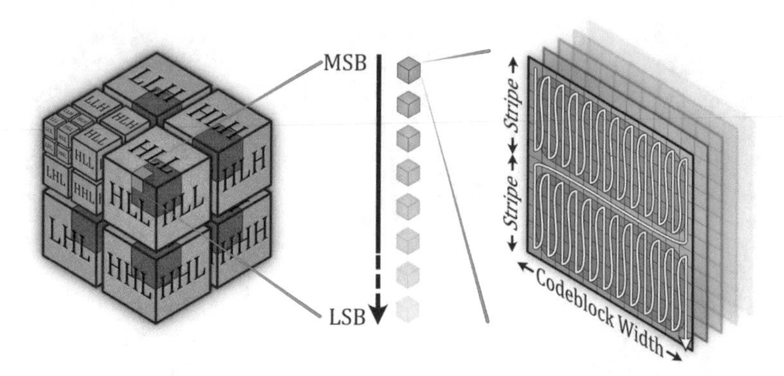


**Fig. 3** 2-dimensional representation of geometric operations performed during Embedded Block Coding with Optimized Truncation stage. Red squares represent a precinct, blue squares a code-block

$$
\begin{aligned}
y(2n+1) &\leftarrow x(2n+1) + (\alpha \times |x(2n) + x(2n+2)|),\\
y(2n) &\leftarrow x(2n) + (\beta \times |y(2n-1) + y(2n+1)|),\\
y(2n+1) &\leftarrow y(2n+1) + (\gamma \times |y(2n) + y(2n+2)|),\\
y(2n) &\leftarrow y(2n) + (\delta \times |y(2n-1) + y(2n+1)|),\\
y(2n+1) &\leftarrow -K \times y(2n+1),\\
y(2n) &\leftarrow (1/K) \times y(2n).
\end{aligned} \tag{1}
$$

In this context, $\alpha = 1.59$ and $\gamma = 0.88$ are the coefficients for the first and second predictor stage. The coefficients $\beta = 0.053$ and $\delta = 0.44$ define the first and second update stage. The dual normalization step is described by the coefficient $K = 1.23$ [9]. Since the discrete-wavelet-transform is a one-dimensional transformation defined for unbounded signals, we can extend the transformation stage to volumetric data-sets by applying the analysis filter-bank along each spatial dimension successively. For bounded data-sets, the undefined samples outside of the finite-length signal need to be related to values inside the signal segment. For odd-length filter taps, this is achieved by applying a whole-point symmetric extend on the signal boundaries [11].

**Entropy Encoding** After the transformation stage, each wavelet sub-band is independently encoded using the EBCOT algorithm described in the JPEG 2000 standard. First, the wavelet sub-bands are rounded down to the nearest integer and subdivided into non-overlapping cubes (see Fig. 3). Spatially related rectangles from the 7 high-bands belonging to the same decomposition level form a precinct. Each precinct is further divided into $32 \times 32 \times 32$ code-blocks, which form the input signal of the entropy encoding stage. These code-blocks are then split into their respective bit-plane fields. The bit-planes are scanned in a zig-zag pattern—from the most significant bit-plane (MSB) to the least significant bit-plane (LSB)—and encoded using three distinct coding passes:

*The significance propagation pass* will encode the current bit of a coefficient if one of its neighbors has become significant in a previous coding pass. To this end, the bit is first assigned to one of nine context labels based on the current sub-band as well as the number of its significant neighbors. The context information, alongside the current bit, is then send to the arithmetic MQ-Coder.

*The refinement pass*, on the other hand, will encode the bits for all coefficients that have become significant in a previous coding pass. Here, the context is assigned according to the significance state of the horizontal, vertical and diagonal direction. Similar to the significant propagation pass, the context label is then passed on to the arithmetic encoder to code the current bit.

*The cleanup pass* encodes all the bits that have not been encoded in the significant propagation and refinement pass. Run coding and context labels are used to encode the string of leftover bits.

During the coding operation, every bit of a bit-plane is only encoded once in one of the three coding passes. The sign bit is encoded as soon as a coefficient becomes significant for the first time. In a post-processing operation, the encoded bit-stream is subdivided into so-called quality layers. These quality layers represent optimized bit-stream truncation points that minimize the distortion for a specific bit-rate. Each successive quality layer will monotonically improve the quality of the data-set reconstruction. The encoded information for each code-block is distributed across all layers. Rate control is handled by defining quality layer 0 to be rate-distortion optimized for the specified compression ratio.

# 4 Results

A comparison between the BigWhoop library and the ZFP and 7-Zip encoders was carried out using a numerical simulation of a Taylor-Green vortex (TGV) decay at $Re = 1600$ (see Fig. 4). The Taylor-Green vortex represents a simple and well-defined hydrodynamics problem, characterized by its turbulent energy cascade. The flow-field is initialized with an analytical solution containing a single length scale. The initial vortex quickly transitions into fully-turbulent dynamics, generating vortices with increasingly smaller turbulent length scales until the turbulent energy is dissipated through viscose effects [5]. The broad turbulent scale spectrum of this test case, at different non-dimensional times t, allowed us to study the effects of length scale on the performance of the individual compressors. The evaluation was performed for $t = 0, 2.5, 5, 7.5, 10, 12.5, 15, 17.5$ and 20. Each time-step contains the density $\rho$, impulses $\rho u$, $\rho v$ and $\rho w$ and energy $\rho E$ on a $nX \times nY \times nZ = 260 \times 260 \times 260$ grid. The file size of one time-step measures 703.052.304 bytes.

To assess the overall quality of the decompressed files we used the peak signal-to-noise ratio metric (PSNR), which is evaluated based on the mean-square-error (MSE):

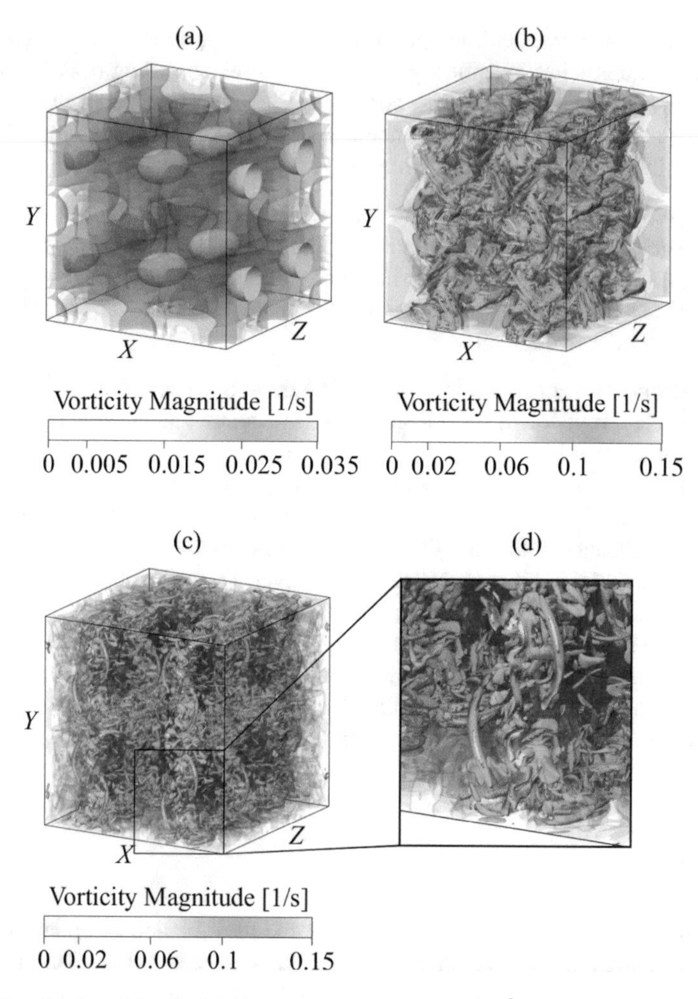


**Fig. 4** Visualization of the Taylor-Green vortex test case on a $256^3$ grid. The vorticity magnitude of the decaying vortex is shown from top left to bottom left for the non-dimensional times $t = 0, 7.5, 15$. [5]

$$MSE = \frac{1}{ijk}\sum_{x=1}^{i}\sum_{y=1}^{j}\sum_{z=1}^{k}|I(x,y,z) - I'(x,y,z)|, \tag{2}$$

$$PSNR = 20\log_{10}\left(\frac{\max(I(x,x,z)) - \min(I(x,x,z))}{\sqrt{MSE}}\right). \tag{3}$$

where $I(x, y, z)$ represents the original, $I(x, y, z)$ the decompressed value and $i, j, k$ the dimension of the volumetric data-set [11]. We found that good reconstruction of the original data-set is typically signaled by a PSNR of $90dB$ or higher.

The absolute error, normalized by the dynamic range of its specific flow-field parameter, was used to evaluate the localized effects of a lossy compression stage.

**Table 1** Comparison of compression ratio, compression time and peak signal-to-noise ratio for the compressed Taylor-Green vortex decay test case using the BigWhoop (BW), ZFP and 7-Zip compressor. The best results appear in bold

| Simulation time | Compression ratio | | | Compression time | | | PSNR | | |
|---|---|---|---|---|---|---|---|---|---|
| | BW | ZFP | 7-Zip | BW | ZFP | 7-Zip | BW | ZFP | 7-Zip[a] |
| 0 | **64.7** | 65 | 41.7 | 6.9 | **1.7** | 114.9 | **133.9** | 61.21 | $\infty$ |
| 2.5 | 62.2 | **62.1** | 6.0 | 4.4 | **2.0** | 160.4 | **123.6** | 60.02 | $\infty$ |
| 5 | **61.6** | 61.1 | 5.8 | 7.2 | **1.9** | 173.2 | **92.0** | 48.16 | $\infty$ |
| 7.5 | 61.0 | **61.1** | 5.6 | 8.8 | **1.9** | 164.0 | **73.8** | 44.26 | $\infty$ |
| 10 | 60.8 | **61.1** | 5.1 | 11.0 | **1.8** | 172.9 | **62.6** | 37.83 | $\infty$ |
| 12.5 | 60.8 | **61.1** | 4.6 | 11.4 | **1.8** | 172.8 | **63.3** | 40.64 | $\infty$ |
| 15 | 60.8 | **61.1** | 4.2 | 11.1 | **1.8** | 186.4 | **64.1** | 40.48 | $\infty$ |
| 17.5 | 60.8 | **61.1** | 3.9 | 10.7 | **1.8** | 183.9 | **65.7** | 43.0 | $\infty$ |
| 20 | 60.9 | **61.1** | 3.7 | 11.1 | **1.9** | 181.5 | **67.6** | 45.2 | $\infty$ |

[a]The PSNR for the 7-Zip compressor is always infinity due to its lossless nature

To assess how much energy is artificially dissipated due to lossy compression, we used the temporal evolution of the energy dissipation rate:

$$\epsilon = \frac{\mu}{\rho_0^2\,\Omega} \int_\Omega \rho\,(\omega \cdot \omega)\,d\Omega. \tag{4}$$

Here, $\mu$ represents the dynamic shear viscosity, $\rho_0$ the density for the initial flow-field state, $\Omega$ the fluid-flow domain and $\omega$ the vorticity vector. The dissipation rate $\epsilon$ is related to the kinetic energy dissipated through small scale structures [13]. It is therefore directly related to high-frequency information present in our Taylor-Green vortex test case, making it highly susceptible to aggressive bit-stream truncation. Our experiments were run on an Intel Core i7-6700 processor with 3.40 GHz and 32 GB of 2133 MHz DDR4 RAM.

The compression results for the transient simulation are listed in Table 1. Compression ratio, compression time and peak signal-to-noise ratio (PSNR) are shown for the BigWhoop, ZFP and 7-Zip encoders. For the initial flow-field (t = 0) all compressors offer good file size reduction. Both the BigWhoop and ZFP algorithm are capable of maintaining the predefined compression ratios throughout the entire time series. However, the 7-Zip encoder is unable to provide significant size reduction for advanced time-steps. This can be attributed to 7-Zip's lossless nature and its inability to exploit spatially correlated information.

The smallest compression time is achieved by the ZFP compressor, followed by BigWhoop and the 7-Zip encoder. The difference in execution speed between the lossy compressors is largely due to the computational complexity of their respective entropy encoders. ZFP's embedded block coding stage simply truncates the bit-stream once the required bit-rate for the local block has been achieved. BigWhoop's EBCOT algorithm, on the other hand, uses a post-processing operation that evaluates a rate-distortion optimized truncation point. This, however, means that bit-planes are compressed and processed that ultimately do not end up in the final code-stream. This

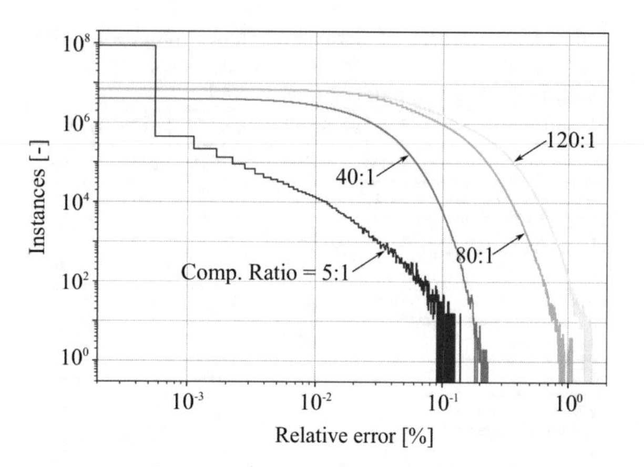


**Fig. 5** Histogram of relative error for the Taylor-Green vortex evolution compressed with the BigWhoop library. The compression ratios 5 : 1, 40 : 1, 80 : 1 and 120 : 1 are shown for the non-dimensional time-step 10

computational overhead should be addressed in the next iteration of the compression library to reduce the overall compression time.

The apparent computational complexity of the BigWhoop library, however, is compensated for by the excellent reconstruction that can be observed throughout the entire time series. In contrast, the infinite support of the discrete cosine transform and the minimalist entropy encoding stage significantly reduce the PSNR that is achievable with the ZFP codec. The clear winner in this category is the 7-Zip encoder with perfect reconstruction throughout. The small compression ratios and excessive compression time, however, limits its usefulness for large-scale CFD simulations.

Before we further analyze the effects of lossy compression on the transient vortex structure of our test case, we first need to establish which compression ratio yields a reconstruction that is still acceptable for flow-field visualization and turbulence statistics. An error of 1% was determined to be reasonable for our applications. The investigation was carried out using the time-step that exhibited the smallest PSNR in our comparative study (see Table 1). Thus, the BigWhoop library was applied to the simulation file for t = 10 with a compression ratio of 5:1, 40:1, 80:1 and 120:1. The respective histograms for the relative error are shown in Fig. 5.

The mean relative error for all compression ratios remain below 0.1%. As expected, the histograms shift towards higher values for an increase in compression ratio. The biggest jump in relative error can be observed between the compression ratios 40:1 and 80:1. The smallest maximum error of 0.23% occurred for a size reduction of 5:1, the largest maximum error of 1.4% for a size reduction of 120:1. Since it yielded good reconstruction with small relative errors throughout, we chose a compression ratio of 120:1 for our investigation.

To analyze the amount of energy lost during compression, the BigWhoop and ZFP compressors were applied to the time series of the Taylor-Green vortex test case at a compression ratio of 120:1. The energy dissipation rate was evaluated for each time-step and compared with the energy dissipation rate of the original data-set (see Fig. 6).

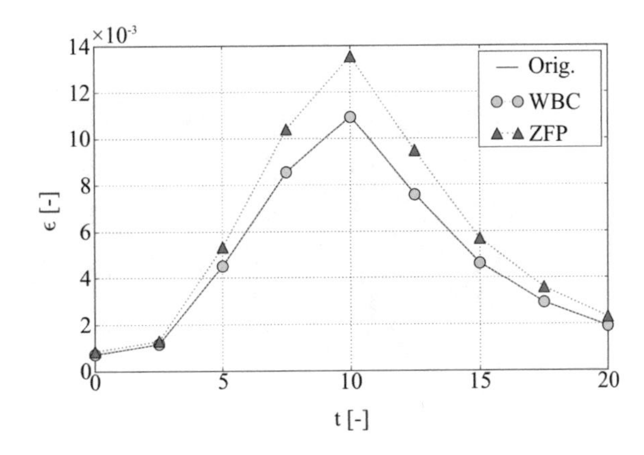


**Fig. 6** Comparison of normalized energy dissipation rate as a function of Taylor-Green vortex evolution for ZFP and BigWhoop encoders and a compression ratio of 120 : 1

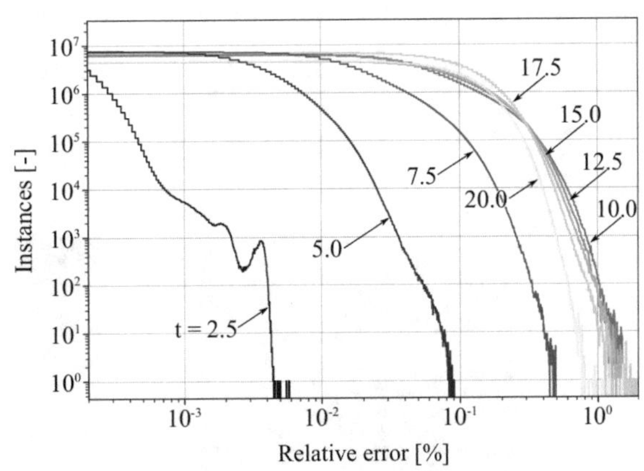


**Fig. 7** Histogram of relative error for the Taylor-Green vortex evolution compressed with the BigWhoop library. The non-dimensional times $t = 2.5, 5, 7.5, 10,$ $12.5, 15, 17.5$ and $20$ are shown for a compression ratio of 120 : 1

Despite the aggressive size reduction applied to the test case, the BigWhoop library is capable of retaining most of the energy contained in the original data-sets. In contrast, the ZFP library artificially dissipates large portions of the internal energy at high compression ratios. This would limit the useful compression ratio available to simulations that rely on the information stored in these small scale structures.

To gain insight into the local distortion introduced by the BigWhoop library, we evaluated the relative error for the transient vortex decay. The histograms shown in Fig. 7 mirror the evolution of the energy dissipation found in the original data-set. Both the maximum ($5 \times 10^{-3}$ %) and mean relative ($8 \times 10^{-5}$ %) error remain negligibly small while the vortex decay is in its initial stages. Most of the turbulent energy is stored in larger length scales that are easy to compress. However, the error histogram is shifted to higher values with an increase in the energy dissipation rate. The largest deviation can be observed at the $t = 10$ with a maximum relative error of 2% and a mean relative error of $8 \times 10^{-2}$ %. Here, the vortex decay has advanced to the point where most of the turbulent energy is stored in small length scales that are dissipated through viscose effects. The high-frequency information these small length scales incur will reduce the overall compression performance. This, in

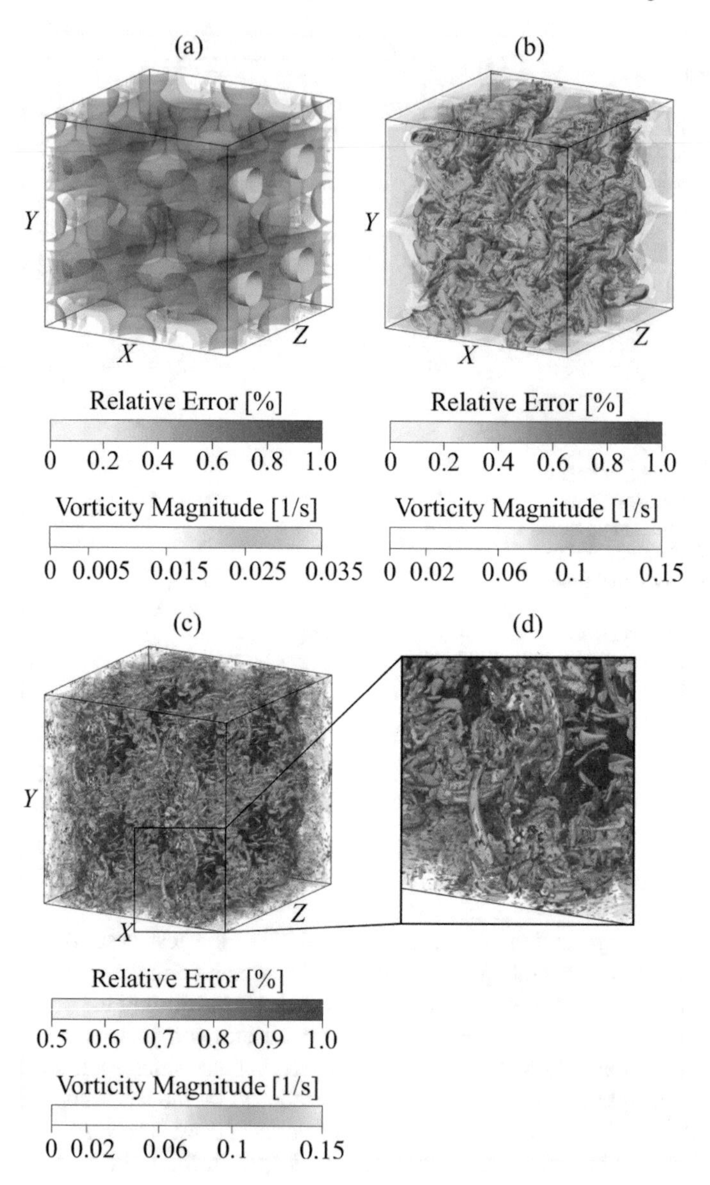


**Fig. 8** Visualization of the compressed Taylor-Green vortex test case on a $256^3$ grid fora compression ratio of 120 : 1. The vorticity magnitude of the decaying vortex is shown from top left to bottom left for the non-dimensional times $t = 0, 7.5, 15$. Volume rendering of the relative error is used to indicate regions of large, compression-induced distortion. [5]

turn, requires a more aggressive truncation of the bit-stream to reach the specified compression ratio. Towards the end of the vortex decay, most of the turbulent energy has already been dissipated from the flow-field domain and thus an improvement in compression performance can be observed. For all time steps, the upper limit of the maximum relative error remains at 2%.

To evaluate where the distortion is introduced, we compressed the three distinct time-steps shown in Fig. 4 with a compression-ratio of 120:1. The relative error for the vorticity magnitude was evaluated for each data sample and normalized with the dynamic range of the vorticity magnitude. Figure 8 shows the decompressed, non-dimensional time-steps t = 0, 7.5 and 15. Isosurfaces for the relative error are used to indicate regions of large, compression-induced distortion. Both time-step t = 0 as well as time-step t = 7.5 show a reasonably well-reconstructed data-set with marginal deviation from the original file. Most of the information is still stored in large length-scales and can thus easily be decorrelated by the discrete wavelet transform. For time-step t = 15 we can clearly observe areas of large, relative errors present in regions defined by small-scale vortex structures. These small-scale fluid-flow structures are synonymous with high-frequency information, which will accumulate in the high-frequency wavelet bands.

This leads to a decrease in compression performance that has to be compensated for by significant truncation of the compressed bit-stream, increasing the local, relative error.

## 5 Conclusion

We presented a wavelet-based lossy compression scheme that allows for the compression of large volumetric floating-point arrays. The proposed technique is based on the JPEG 2000 algorithm [11]. A comparison with established compression techniques was done using three-dimensional data-sets from a numerical simulation of a Taylor-Green vortex decay. Based on the results of our study we found that our compression approach can significantly decrease the overall flow-field distortion while maintaining a high compression ratio. The results, however, also show the need for a more optimized implementation of our wavelet-based codec. Looking ahead, we will seek further improvements in data throughput by using OpenMP and MPI. Furthermore, we hope to extend our algorithm to numerical simulations that employ mesh refinement techniques. The effects of our lossy wavelet-based compressor on boundary layer flows will need to be assessed.

**Funding**: This work was supported by a European Commission Horizon 2020 project grant entitled EXCELLERAT [Grant agreement ID: 823691].

**Acknowledgements** The authors would like to thank Björn Selent, Christoph Wenzel, Duncan Ohno and Gabriel Axtmann, at the University of Stuttgart's Institute of Aerodynamics and Gasdynamics, for providing valuable insight into fluid flow mechanics and turbulence statistics.

## References


1. Acharya, T., Tsai, P.: JPEG2000 Standard for Image Compression: Concepts. Algorithms and VLSI Architectures. Wiley, New Jersey (2005)
2. Bruylants, T., Munteanu, A., Schelkens, P.: Wavelet based volumetric medical image compression. Signal Process.: Image Commun. **31**, 112–133 (2015). https://doi.org/10.1016/j.image.2014.12.007
3. Christopoulos, C., Skodras, A., Ebrahimi, T.: The JPEG2000 still image coding system: an overview. IEEE Trans. Consum. Electron. **46**(4), 1103–1127 (2000). https://doi.org/10.1109/30.920468
4. Gamito, M.N., Salles Dias, M. Tescher, A.G.: Lossless coding of floating point data with JPEG 2000 Part 10. Appl. Digit. Image Process. XXVII **5558**, 276–287 (2004). https://doi.org/10.1117/12.564830
5. Jacobs, T., Jammy, S.P., Sandham, N.D.: OpenSBLI: a framework for the automated derivation and parallel execution of finite difference solvers on a range of computer architectures. J. Comput. Sci. **18**, 12–23 (2017). https://doi.org/10.1016/j.jocs.2016.11.001
6. Lindstrom, P.: Fixed-rate compressed floating-point arrays. IEEE Trans. Vis. Comput. Graph. **20**(12), 2674–2683 (2014). https://doi.org/10.1109/TVCG.2014.2346458
7. Lindstrom, P., Isenburg, M.: Fast and efficient compression of floating-point data. IEEE Trans. Vis. Comput. Graph. **12**(5), 1245–1250 (2006). https://doi.org/10.1109/TVCG.2006.143
8. Loddoch, A., Schmalzl, J.: Variable quality compression of fluid dynamical data sets using a 3-D DCT technique. Geochem., Geophys., Geosystems **7**(1), 1–13 (2006). https://doi.org/10.1029/2005GC001017
9. Rabbani, M., Joshi, R.: An overview of the JPEG 2000 still image compression standard. Signal Process.: Image Commun. **17**(1), 3–48 (2002). https://doi.org/10.1016/S0923-5965(01)00024-8
10. Schmalzl, J.: Using standard image compression algorithms to store data from computational fluid dynamics. Comput. Geosci. **29**, 1021–1031 (2003). https://doi.org/10.1016/S0098-3004(03)00098-0
11. Taubman, D., Marcellin, M.: JPEG2000: Image Compression Fundamentals. Standards and Practice. Springer, US, New York City (2002)
12. Usevitch, B.E.: JPEG2000 compatible lossless coding of floating-point data. EURASIP J. Image Video Process. textbf(2007), 1–8 (2007). https://doi.org/10.1155/2007/85385
13. Welch, T.A.: The dynamics of enstrophy transfer in two-dimensional hydrodynamics. Phys. D: Nonlinear Phenom. **48**(2–3), 273–294 (1991). https://doi.org/10.1016/0167-2789(91)90088-Q
14. Welch, T.A.: A technique for high-performance data compression. Computer **17**(6), 8–19 (1984). https://doi.org/10.1109/MC.1984.1659158
15. ARM Developer Suite, N.N: AXD and ARMSD Debuggers Guide. ARM, Cambridge (2001)

# A Method for Stream Data Analysis

**Li Zhong**

**Abstract** Due to the recent advances in hardware and software, a number of applications generate a huge amount of data at a great velocity which make big data stream become ubiquitous. Different from static data analysis, processing data stream imposes new challenges for algorithms and methods which need to incrementally deal with incoming data with limited memory and time. Furthermore, due to the inherent dynamic characteristics of streaming data, algorithms are often required to solve problems like concept drift, temporal dependencies, load imbalance, etc. In this paper, we discuss state of the art researches on data stream analysis which employed rigorous and methodical approaches, especially deep learning. Besides, a new method for processing data stream based on the latest development of GAN is proposed. And finally the future work to be done is discussed.

## 1 Introduction

In the past few years, we have witnessed a fundamental shift from parsimonious, infrequent measurement to nearly continuous monitoring and analyzing in a lot of areas, including geography science, remote sensing, healthcare, finance and so on. Rapid advances in the areas stated above have generated a rapid growth in the size and complexity of data stream archives. A data stream is an ordered sequence of items that arrives in timely order. Different from data in traditional static databases, data stream usually has the following features [16]:

1. Data arrives continuously
2. The size of the data is unbounded
3. Data usually needs to be discarded after processing due to the hardware limitation
4. The data generation is non-stationary, and its probability distribution always changes over time
5. The data usually come with high speed

L. Zhong (✉)
HLRS, Nobelstr. 19, Stuttgart, Germany
e-mail: li.zhong@hlrs.de



M. M. Resch et al. (eds.), *Sustained Simulation Performance 2019 and 2020*,
https://doi.org/10.1007/978-3-030-68049-7_8

In most cases, data streams are far too large to fit in memory and thus are often required to process in a near real-time manner. From this stand-point, performing linear scans of the data and analysis only on a portion of the data seems to be the only acceptable method. However, it is obvious that extracting potential useful knowledge from data stream is quite challenging with this method. What makes this become a more difficult problem is that the experience we gained from batch data analysis can barely help much here. In traditional batch data analysis techniques, it is assumed that there is a finite amount of data with stationary probability distribution which could be processed without any time and step limitation. Taken the natural features of data stream into consideration, traditional batch data analysis techniques are not sufficient in such scenarios. The idea of streaming analysis is that each of the received data slice is processed in the dedicated data processing node. Such processing includes removing duplicates, filling missing data, data normalization, parsing, feature extraction, which are typically done in a single pass due to the high data rates of external feeds [20]. When a new data slice arrives, this node is triggered, and it expels the data that are older than the time specified. Most of the data generated in a real-time data stream need real-time data analysis. In addition, the output must be generated with low-latency and any incoming data must be reflected in the newly generated output within seconds. As a result, there is a crucial need for parallel architectures and scalable computing platforms. The difference between the requirements of batch data analysis and stream data analysis is listed in Table 1.

In addition, data streams pose several challenges for analysis algorithms, including, but not limited to, concept drifts [28], temporal dependencies [32], massive amount of instances, limited labeled instances, novel classes, feature drifts [2, 31], and restricted resources (time and memory) requirements. On top of that, problems found in a batch-learning setting are also present in a data stream context, for example, absent values, overfitting, noise, irrelevant features, class imbalance, and others [14]. Thus, while data stream analysis has been studied extensively in the past, its importance only continues to grow. This paper will discuss a method which could potentially solve the open research problems.

**Table 1** Comparison of batch data analysis and stream data analysis

| | Batch data analysis | Stream data analysis |
|---|---|---|
| Data feature | Data chunks | Data flow |
| Data size | Stationary and finite | Dynamic and infinite |
| Data storage | Store | No Store or store a portion |
| Processing limitation | – | Few passes over a portion of data |
| Processing time requirement | No special requirement | Seconds or milliseconds |

## 2 Related Work

In the past several years, deep learning-based methods have gained much popularity with their promising performance in data stream analysis [31]. For instance, the AutoEncoder (AE) [33] is a popular deep learning model by inspecting its reconstruction errors in data stream analysis. Others like Deep Autoencoding Gaussian Mixture Model (DAGMM) [33] and LSTM Encoder-Decoder [10] have also reported good performance. Based on this, a lot of research has been done in data stream analysis applying deep neural networks. For example, in the work done by Calandra et al. [8], Deep Belief Networks have been proposed to generate the samples from the training data distribution, which are used at a later stage for retraining and classification.

In a more recent work conducted by Rusu et al. [25], Progressive Networks are designed for effective knowledge transfer across multiple sequentially arriving tasks. In the proposed approach, a single deep neural network is initialized and trained for the first given task. Each time a new task is added, a new neural network with the same architecture as previous ones is initialized. At the same time, directed connections from all previous models to the current models are initialized and serve as knowledge transfer functions.

Pathnet, as another method is proposed by Banarse et al. [2]. It can be considered as an evolution of Progressive Networks, is represented by a huge Neural Network with embedded agents, that are evolving to find the best matching paths through the network for a current task. After the task is learned and the new one is introduced, parameters of the sub-network containing the optimal pathway are "freezed" not to be modified by back-propagation when training for new tasks.

The main problem is that most of the Deep Learning methods are prone to forgetting the concepts that are no longer represented by the dataset they are trained on. In literature this phenomenon is known as catastrophic forgetting [23]. At the same time, training Neural Networks is based on gradient backpropagation, which is slow and often requires passing through the dataset many times [4]. Besides, most of the data are unlabeled, supervised machine learning methods are unable to exploit the large amounts of data due to the lack of labled data. On the other hand, current unsupervised machine learning approaches have not fully exploited the spatial-temporal correlation and other dependencies amongst the multiple variables in the system for data stream analysis.

## 3 Prelimitary

With the rapid development of deep learning techniques, briefly three deep neural network architectures are reviewed, based on which a method is proposed. The AutoEncoder (AE) [33] is a popular deep learning model by inspecting its reconstruction errors in data stream analysis. Another popular method, Generative Adversarial Network (GANs) [15], continue to show increasing power in synthesizing highly realistic observations, and have found wide applicability in various fields. In particular for time varying data stream, the Long Short-Term Memory(LSTM) architecture [18], which overcomes some modeling weaknesses of RNNs, is conceptually attractive. In this section, we briefly review these three above mentioned building blocks on which our method is constructed.

### 3.1 Generative Adversarial Networks (GANs)

Recently, the Generative Adversarial Networks (GAN) framework has been proposed to build generative deep learning models via adversarial training [24]. It continues to show its powerful capability for unsupervised tasks in various fields [1, 21], especially in generating realistic-looking data [7, 11]. Unlike traditional methods, the GAN-trained discriminator learns to detect fake data from real data in an unsupervised fashion, and features from the discriminator or inference net could improve performance of classifiers when limited labeled data is available [24], making it an attractive unsupervised machine learning technique for data analysis. GANs were proposed by Goodfellow et al. [15], who explained the theory of GANs learning based on a game theoretic scenario. In the scenario of data stream, where always exist the limitation of memory and processing times, GAN has the potential to solve this problem through realistic-looking data generation.

A GAN is composed of two main parts, a generator G and a discriminator D. The objective of the simultaneous training of the generator and discriminator in an adversarial fashion is:

$$\min_{G} \max_{D} \mathbb{E}_{x \sim q_{data}(x)}[log D(x)] + \mathbb{E}_{z \sim p(z)}[log(1 - D(G(z)))] \tag{1}$$

where $p(z)$ is a simple distribution and $q_{data}(x)$ is the underlying data distribution from which we observe samples. A typical GAN architecture is illustrated as in Fig. 1.

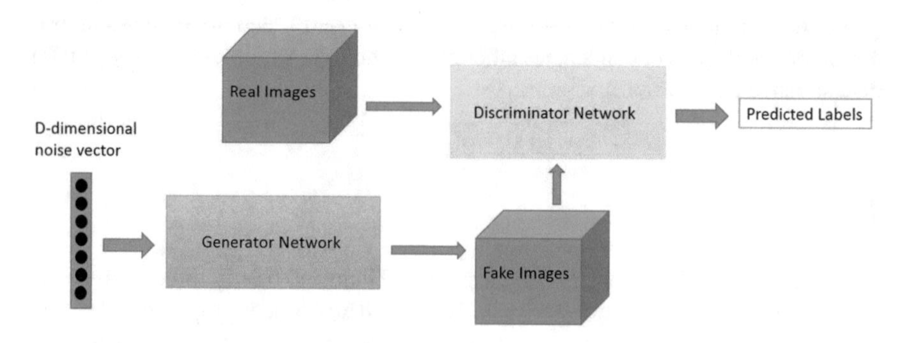


**Fig. 1** Illustration of generative adversarial model

### 3.2 AutoEncoder

An AutoEncoder is a neural network model composed of two main parts: an encoder and a decoder. The encoder part maps the input into a vector representation, and the decoder part reconstructs the original input through the vector representation as the output. Except from the most frequently use scenarios like dimension reduction and feature learning, AutoEncoder is becoming more and more widely used for learning generative models of data [19]. Variational AutoEncoder(VAE) [19] as an famous generative model can model the distributed of the data also through an encoder-decoder structure. The principle is a simple distribution with known parameters and super imposable characteristics can theoretically fit any distribution by combining with neural networks [30]. A typical AutoEncoder architecture is illustrated as in Fig. 2.

### 3.3 Long Short Term Memory Recurrent Neural Network (LSTM-RNN)

Long Short-Term Memory (LSTM) is a specific recurrent neural network (RNN) architecture that was designed to model temporal sequences and their long-range dependencies more accurately than conventional RNNs [26]. The capability of learning information in backward or forward time steps of LSTM is implemented through a special unit called memory blocks. The memory block is consisted of two main

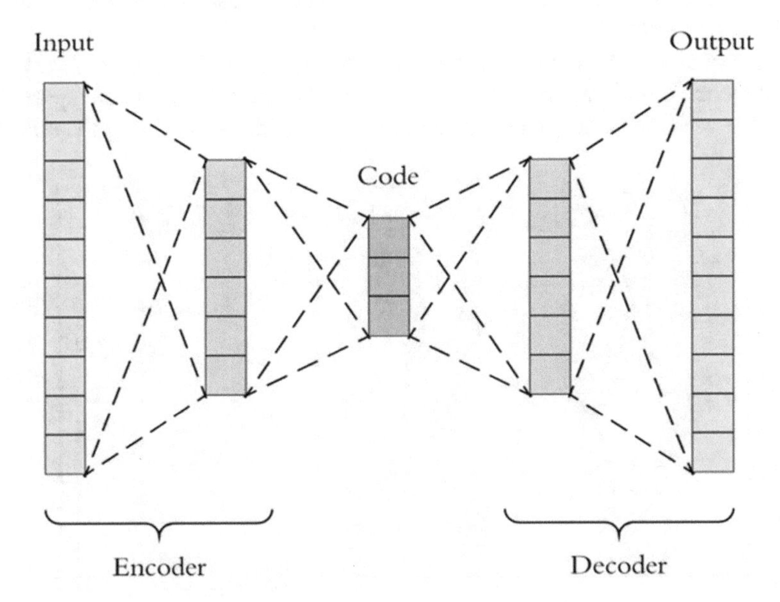


**Fig. 2** Illustration of autoencoder model

parts: a self-connected memory cell to store the temporal state of the network and a special multiplicative unit namely 'gate' to control the flow of the information. And each memory block has an input gate which controls the flow of the input activation into memory cells and an output gate which controls the output flow of cell activations into the rest of the network. In addition, there is also a forget gate in memory block, which scales the internal state of the cell before adding it as input to the cell through the self-recurrent connection of the cell, therefore adaptively forgetting or resetting the cells memory [12]. In addition, the modern LSTM architecture contains peephole connections from its internal cells to the gates in the same cell to learn precise timing of the outputs [13].

# 4 Proposed Method

As described in Sect. 3, an AutoEnocer network has shown its advantage in data compression and feature extraction. And a GAN network has a great performance in realistic-look data generation. It can adapt to data classes which are never seen by the model before, while preserving previously learned information. More specifically, generative models are able to regenerate the historical training data that we didn't keep [22]. But with AutoEncoder and GAN, the dependencies among the incoming data slices are abandoned. However, in most cases of data stream analysis the dataset cannot be regarded as i.i.d., and the potential temporary dependencies among the data flow must be taken into consideration. Therefore, a method which combines GAN, AutoEncoder and LSTM is proposed, where the LSTM is used as the encoder, the generator and the discriminator, so that this method could benefit from the advantages of all three models.

The proposed network architecture is depicted as Fig. 3. The input to the encoder is a sequence of data which will be encoded to the vector in the latent space. The generator uses the vector in the latent space to generate the vector in the real-time

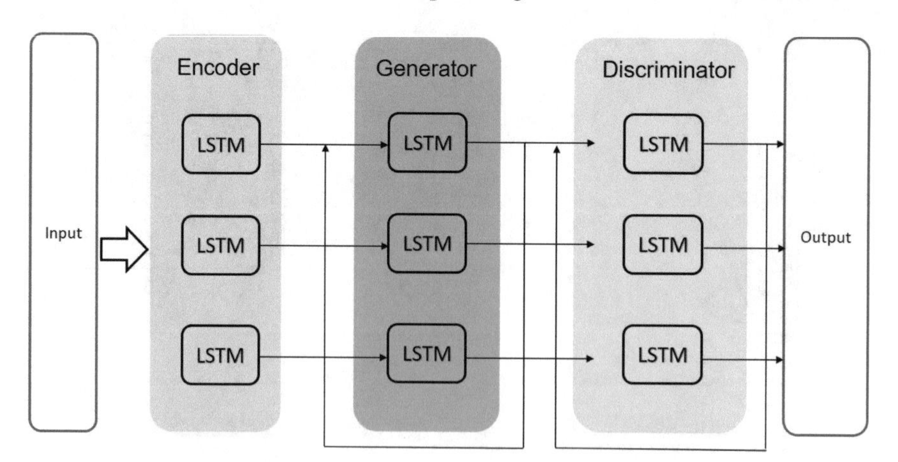


**Fig. 3** Illustration of proposed new network structure

space, and the discriminator outputs a vector which indicates whether the vector generated by the generator obeys the distribution of the input data. To alleviate the catastrophic forgetting issue, generative replay (or pseudo-rehearsal) [29] as an effective and general strategy for lifelong learning is adopted here by both the generator and the discriminator. Generative replay has advantages in addressing privacy concerns and remembering potentially more complete data information [9].

Therefore, this proposed method need not to store and reuse previous observations. What is more, the efficiency is improved, as training could be accelerated greatly by divising better methods for coordinating Generator and Discriminator or determining better distributions to sample during training.

## 5 Conclusion and Outlook

In this paper, a method based on GAN, AutoEncoder and LSTM is proposed to solve the problems in data stream analysis, like catastrophic forgetting, anomaly detection and concept drift. The new algorithm can be proved by manifold use cases in many areas, for example, Internet of Things (IoT), wearable health devices, Smart City, etc. In the era of IoT, an enormous amount of sensing devices collects and/or generate various sensory data over time for a wide range of fields and applications [11]. Based on the nature of the application, these devices will result in big or fast/real-time data stream [17]. Applying the new analytic methods over such data streams to discover new information, predict future insights, and make control decisions is a crucial process that makes IoT a worthy paradigm for businesses and a quality-of-life improving technology.

As analyzing Big Data in a timely manner necessitates a high performance and scalable computing infrastructure [6], especially in the era of edge computing, where a distributing environment that automatically streams partial data streams to a global centre when local resources become insufficient is required, HPC continues to play an important role. Thus research efforts should be geared towards scalable algorithms and methods that will utilize the HPC infrastructure to benefit from the the computing mode, resource allocation and parallelization strategies. Consequently, a data stream analysis tool based on the novel method which provides high performance analysis will be developed in the future.

## References

1. Ak, K., Lim, J., Tham, J., Kassim, A.: Attribute manipulation generative adversarial networks for fashion images. In: ICCV, pp. 10541–10550 (2019)
2. Banarse, F.D., Blundell, C., Zwols, Y., Ha, D., Rusu, A.A., Pritzel, A., Wierstra, D.: Pathnet: Evolution Channels Gradient Descent in Super Neural Networks (2017). arXiv preprint arXiv:1701.08734
3. Barddal, J. P., Gomes, H. M, Enembreck, F., Pfahringer, B.: A survey on feature drift adaptation: definition, benchmark, challenges and future directions. J. Syst. Softw. (2016)

4. Besedin, A., Blanchart, P., Crucianu, M., Ferecatu, M.: Deep online storage-free learning on unordered image streams. In: Joint European Conference on Machine Learning and Knowledge Discovery in Databases (pp. 103–112). Springer, Cham (2018)
5. Besedin, A., Blanchart, P., Crucianu, M., Ferecatu, M.: Evolutive deep models for online learning on data streams with no storage. In: Workshop on Large-scale Learning from Data Streams in Evolving Environments (2017)
6. Binas, J., Neil, D., Liu, S.-C., Delbruck, T.: DDD17: end-to-end DAVIS driving dataset. In: ICML-17 Workshop on Machine Learning for Autonomous Vehicles. Sydney, Australia (2017)
7. Brock, A., Donahue, J., Simonyan, K.: Large scale GAN training for high fidelity natural image synthesis. In: ICLR (2019)
8. Calandra, R., Raiko, T., Deisenroth, M., Pouzols, F.: Learning deep belief networks from non-stationary streams. Artif. Neural Netw. Mach. Learn. ICANN **2012**, 379–386 (2012)
9. Cong, Y., Zhao, M., Li, J., Wang, S., Carin, L.: GAN Memory with No Forgetting (2020). arXiv preprint arXiv:2006.07543
10. Edan, H., Shabtai, A.: Using LSTM encoder-decoder algorithm for detecting anomalous ads-b messages. Comput. Secur. **78**, (2018)
11. Esteban, C., Hyland, S. L., Rätsch, G.: Real-Valued (Medical) Time Series Generation with Recurrent Conditional Gans (2017). arXiv preprint arXiv:1706.02633
12. Gers, F.A., Schmidhuber, J., Cummins, F.: Learning to forget: Continual prediction with LSTM (1999)
13. Gers, F.A., Schraudolph, N.N., Schmidhuber, J.: Learning precise timing with LSTM recurrent networks. J. Mach. Learn. Res. **3**, 115–143 (2002)
14. Gomes, H. M., Barddal, J. P., Enembreck, F., Bifet, A.: A survey on ensemble learning for data stream classification. ACM Comput. Surv. **50**(2), 36 pages. https://doi.org/10.1145/3054925
15. Goodfellow, I., Pouget-Abadie, J., Mirza, M., Xu, B., Warde-Farley, D., Ozair, S., Bengio, Y.: Generative adversarial nets. In: Advances in Neural Information Processing Systems (pp. 2672–2680) (2014)
16. Guha, S., Koudas, N., Shim, K.: Data-streams and histograms. In: Proceedings of the Thirty-Third Annual ACM Symposium on Theory of Computing (pp. 471–475). ACM (2001)
17. Han Jiawei, J.P., Kamber,M.: Data Mining: Concepts and Techniques. Elsevier
18. Hochreiter, S., Schmidhuber, J.: Long short-term memory. Neural Comput. **9**(8), 1735–1780 (1997)
19. Kingma, D.P., Welling, M.: Auto-encoding variational bayes (2013). arXiv preprint arXiv:1312.6114
20. Kolajo, T., Daramola, O., Adebiyi, A.: Big data stream analysis: a systematic literature review. J. Big Data **6**(1), 47 (2019)
21. Li, D., Chen, D., Goh, J., Ng, S.K.: Anomaly detection with generative adversarial networks for multivariate time series (2018). arXiv preprint arXiv:1809.04758
22. Li, D., Chen, D., Shi, L., Jin, B., Goh, J., Ng, S. K.: MAD-GAN: Multivariate Anomaly Detection for Time Series Data with Generative Adversarial Networks (2019). arXiv preprint arXiv:1901.04997
23. McCloskey, M., Cohen, N.J.: Catastrophic interference in connectionist networks: the sequential learning problem. Psychol. Learn. Motiv. **24**, 109–165 (1989)
24. Mogren, O.: C-RNN-GAN: Continuous Recurrent Neural Networks with Adversarial Training (2016). arXiv preprint arXiv:1611.09904
25. Rusu, A., Rabinowitz, N.C., Desjardins, G., Soyer, H., Kirkpatrick, J., Kavukcuoglu, K., Pascanu, R., Hadsell, R.: Progressive neural networks (2016). arXiv preprint arXiv:1606.04671
26. Sak, H., Senior, A.W., Beaufays, F.: Long short-term memory recurrent neural network architectures for large scale acoustic modeling (2014)
27. Sun, Y., Song, H., Jara, A.J., Bie, R.: Internet of things and big data analytics for smart and connected communities. IEEE Access **4**, 766–773 (2016)
28. Tsymbal, A.: The Problem of Concept Drift: Definitions and Related Work. Technical Report (2004)

29. van de Ven, G.M., Tolias, A.S.: Generative replay with feedback connections as a general strategy for continual learning (2018). arXiv preprint arXiv:1809.10635
30. Zhang, C., Li, S., Zhang, H., Chen, Y.: VELC: A New Variational AutoEncoder Based Model for Time Series Anomaly Detection (2019)
31. Žliobaitė, I., Pechenizkiy, M., Gama, J.: An overview of concept drift applications. In: Big Data Analysis: New Algorithms for a New Society (pp. 91–114). Springer, Cham (2016)
32. Žliobaitė, I., Bifet, A., Read, J., Pfahringer, B., Holmes, G.: Evaluation methods and decision theory for classification of streaming data with temporal dependence. Mach. Learn. **98**(3), 455–482 (2015)
33. Zong, B., Song, Q., Min, M. R., Cheng, W., Lumezanu, C., Cho, D., Chen, H.: Deep autoencoding gaussian mixture model for unsupervised anomaly detection (2018)

# CFD Simulation with Microservices for Geoscience Applications

**Alexey Cheptsov**

**Abstract** The current geoscience applications are confronted by two major challenges—integration with numerous diverse sensor devices and use in real-time use case scenarios. Whilst the challenge of service integration is addressed by the concept of Cyber-Physical systems, which aims to incorporate sensor data in application workflows, the usage of High Performance Computers helps minimize the execution time to fulfill the real time scenarios requirements. However, the existing programming models do not allow scientific workflows to take advantage of both technologies simultaneously. This paper contribution offers an approach to encapsulation of workflow-based applications into services, which are flexible enough to run on heterogeneous, distributed infrastructures spanning over both industrial sensor services and parallel computing systems. The approach is demonstrated on a computational fluid dynamics simulation study of aerodynamic processes in complex underground mine ventilation networks.

## 1 Introduction

Geoscience applications rely largely on simulation, which is used to retrieve and investigate the state of the targeted complex dynamic systems and also to predict their behavior under certain conditions in the future. One of the typical geoscience simulation tasks is served by the Computational Fluid Dynamics (CFD)—a technique that is used to study the behavior of liquids and gases in complex environments. The CFD technique can be used to model many safety-critical processes, as for example the propagation of waves as a result of tsunamis, distribution of volcanic plumes after an eruption, or distribution of air and hazardous gases in underground ventilation objects like coalmines. As all the other CFD applications, these studies are based on complex mathematical models (like Navier-Stokes equation), which generally create a good deal of uncertainty for the simulation results and also require

A. Cheptsov (✉)
High Performance Computing Center Stuttgart (HLRS), University of Stuttgart,
Nobelstraße 19, 70569 Stuttgart, Germany
e-mail: cheptsov@hlrs.de



M. M. Resch et al. (eds.), *Sustained Simulation Performance 2019 and 2020*,
https://doi.org/10.1007/978-3-030-68049-7_9

computationally expensive solution methods. In practice, geoscience applications are often organized in workflows with several interconnected components, each implementing a specific part of the application logic and running on a dedicated resource of the distributed system. The computationally intensive parts of the workflow are usually executed on parallel High Performance Computing (HPC) resources whilst the data acquisition happens on the sensor nodes. However, the workflow approach has several limitations. Firstly, the workflow management software requires quite a rich functionality of resource management, application scheduling, monitoring and other middleware that is related to the workflows execution (like Pegasus, as described by Chang et al. in [3]), which are difficult to provision on the productional HPC systems. Secondly, the workflow-based specification of applications requires quite an extensive metadata schema, which might require substantial change from one execution scenario to the other. Lastly, the implementation of the control flow across the components that include parallelized applications, e.g. with the help of the Message-Passing Interface (MPI), is difficult due to the functional orientation of the latter. In the other words, it is technologically difficult to build a workflow management system that would enable running applications of both types (event-based serial ones, functionally-oriented parallel ones) within the same control and data flow logic and on distributed heterogeneous resources. This paper contribution introduces an alternative approach, which allows MPI applications to be built in a service-oriented way, thus allowing for flexibility of data processing, as required by geoscience applications, whilst keeping a much lower management overhead than in case of traditional workflow management systems. The proposed approach is facilitated by a "Multiple Instruction Multiple Data" (MIMD)-based programming model, which could be inheritably implemented in MPI-parallel applications. The use of the elaborated programming model is illustrated on an implementation of a CFD simulation application for underground coalmine ventilation tasks. The remainder of the paper is organized as follows: Sect. 2 gives an introduction of ventilation networks and simulation tasks for them. Section 3 elaborates a microservices based architecture and methodology for implementation of simulation applications. Section 4 discusses results that are obtained for the evaluation cases. Section 5 concludes the paper and discusses main outcomes.

## 2 Ventilation Networks as Objects of Modelling and Simulation

Fossil coal remains one of the most important energy sources, along with gas, oil, and regenerative energy technologies. In particular, it holds a leading position among the fossil fuels with proven reserves of over 1 Tera-Ton (as shown in [1]) worldwide. At the same time, the coal industry is one of the most dangerous and security-critical industry branches, due to the complexity of the coal mining process from the deep underground locations (up to many tens of kilometers under the surface) and considering the aspects of a high gas content in the obtained coal masses. Ventilation

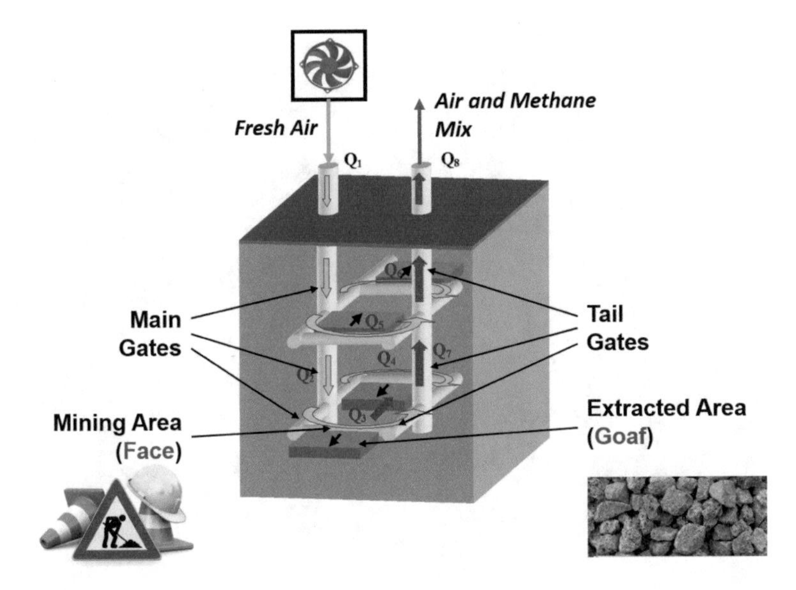


**Fig. 1** Structure of underground mine ventilation

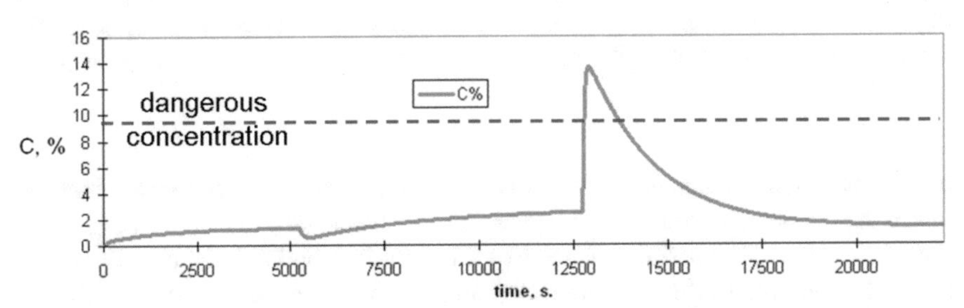


**Fig. 2** Analysis of march gas concentration in time

is the most important aspect of the security provisioning in underground production areas of coalmines (see Fig. 1)—it aims to ensure the underground mining workers with the necessary amount of fresh air and also to dilute the hazardous gases (mainly $CH_4$—marsh gas) that are emitted during the coal loosing, transporting and other technological activities of the mining process.

The degassing procedure is of especial importance for the safety of the mining process—the marsh gas concentration that exceeds the upper threshold can lead to vast exposures in the underground area with major human losses and injuries (see Fig. 2). Historically, gas exposures have been the major reason of big catastrophes that have happened in the coal industry since the beginning of the industrialization era and up to nowadays.

The unpredictable nature of the gas emission is the major challenge for the operation of underground coalmines. Solving the challenge of air and gas distribution

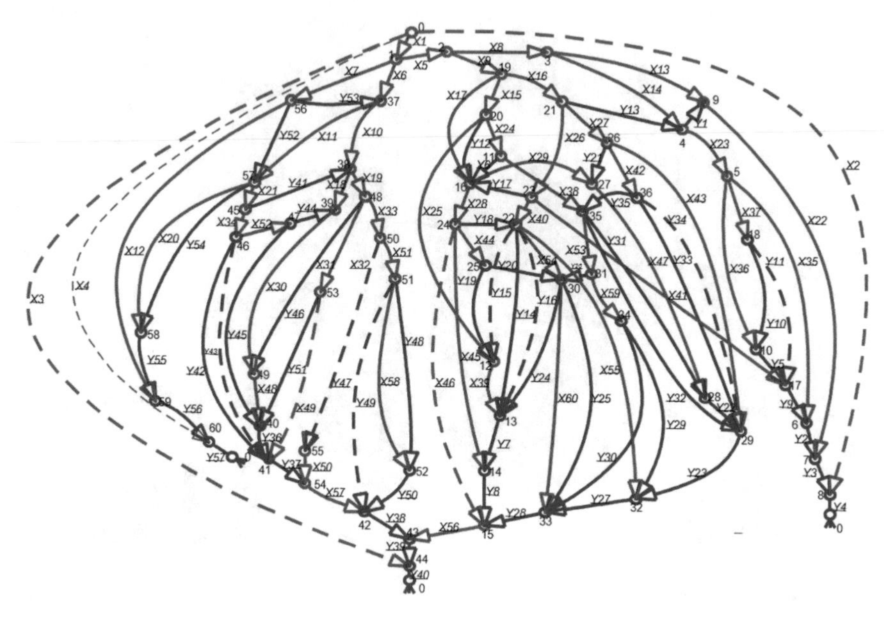

**Fig. 3** Illustration of real-complexity ventilation network with 117 branches and 61 connection nodes (coalmine South-Donbass Nr. 3 in Ukraine). The coloring of airflows is only used for better readability

prediction requires a detailed knowledge of all dynamic processes that are happening in the elements of real industrial ventilation objects (see example in Fig. 3). In most cases, the only possible chance to get insight into optimal planning of air distribution along the ventilation elements and plan the gas dilution actions is the use of modelling and simulation, coupled with the information coming from sensors, which are measuring airflow speed and gas concentration.

Typical CFD models of coalmine ventilation are based on the macroscopic definition of the multiphase flow based on Navier-Stokes equation, e.g., in the following general form for the air distribution in one branch of the ventilation network (e.g., as elaborated by Svjatnjy [8] for a 1-D approximation):

$$\begin{cases} -\frac{\partial \mathrm{P}}{\partial \xi} = -\frac{2\rho}{\mathrm{F}^2}\mathrm{Q}\frac{\partial \mathrm{Q}}{\partial \xi} + \frac{\rho}{\mathrm{F}}\frac{\partial \mathrm{Q}}{\partial \mathrm{t}} + \mathrm{r}\mathrm{Q}^2 + \mathrm{r}'(\mathrm{t})\mathrm{Q}^2 \\ -\frac{\partial \mathrm{P}}{\partial \mathrm{t}} = \frac{\rho \mathrm{a}^2}{\mathrm{F}}\frac{\partial \mathrm{Q}}{\partial \xi} - \frac{\rho \mathrm{a}^2}{\mathrm{F}} q \end{cases} \tag{1}$$

where P—pressure, Q—airflow, t—time, $\xi$—spatial coordinate, and other values—aerodynamic parameters and coefficients.

The gasflow distribution analysis is based on the data obtained from the sensors, which are fed in the transport equation (1) similarly to the airflow (for example, as described by Stewart et al. [7]). Given the insufficient coverage of the underground production areas by sensors, additional prediction models might be used, e.g., as described by the previous publication [5] for the goaf—a mine area that remains

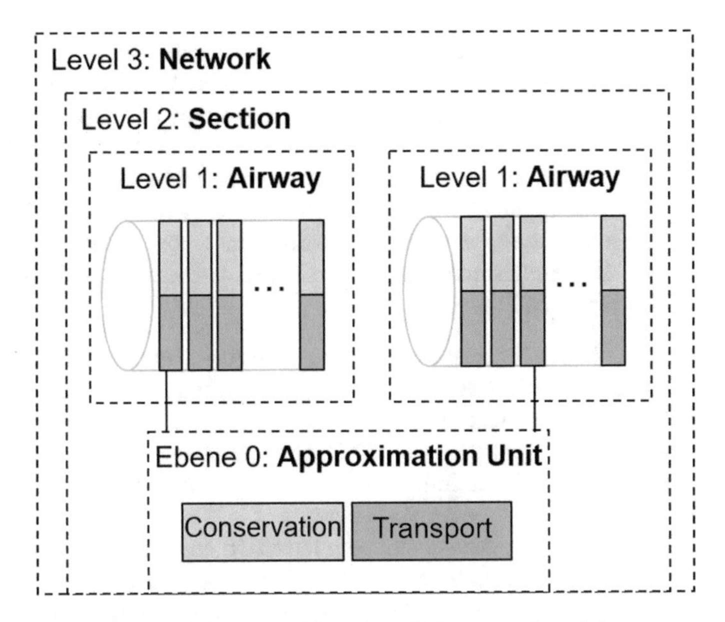


**Fig. 4** Hierarchical approach for composition of ventilation network models

after the coal mining. The model of the whole ventilation network (see example in Fig. 3) is built from the models of each of its elements/branches in the general form (1), extended by the boundary conditions in the nodes and following the following hierarchical organization of the elements: approximation unit of numerical method –> element/airway –> section –> network (see Fig. 4). Further coalmine ventilation simulation tasks include reduction of energy that is required for operation of the main mine fans (which usually consume up to 40% of the overall coalmine energy), as shown by Clausen [6].

Improvement of the quality of the CFD computational models has been the focus of many research activities in the last decade. In particular, lots of activities have been concentrated around integration of sensor data into the simulation process, for example in the form of initial or boundary conditions for the mathematical models, serving the basis for the simulation packages. The "online" sensor data integration allows, among others, to specialize the generic models, i.e., adapt the model parameters to the specifics of the targeted simulation object. With the proliferation of the unified (Ethernet) networking standards (both wired and wireless) in the field of industrial and automated systems (like Industrial Ethernet solutions *PROFINET*, *Modbus*, and others, as described by Kay et al. [1]), sensors have become a vital part of the distributed computing infrastructure and, most essentially, of their software applications. Such an infrastructure which allows a seamless integration of the network-enabled data acquisition devices (such a $CH_4$—marsh gas sensor) with the "traditional" computation and storage facilities is often referred in the literature as Cyber-Physical Systems (CPS). Being initially elaborated for the automotive and industrial automation domains (as described by Broy [2]), the CPS concept is gaining an increasingly growing popularity for the other industrial and scientific areas, including the geoscience applications domain, as targeted by this paper.

Nowadays, geoscience applications are represented by parallel software packages that require large-scale computing and storage facilities of HPC and Cloud infrastructures. Those application are usually developed by means of MPI or other parallelization standards and have a limited ability to incorporate data from external (distributed over the communication network) sensors and other acquisition devices due to the following limitations:

- Heterogeneity of the CPS distributed infrastructure—many sensor devices are provided on the basis of a host system, whose architecture might differ from the typical HPC, Cluster, or Cloud environment but still requires a seamless integration within the distributed application workflows. However, the standard parallelization approaches require a uniform infrastructure with the compute nodes of the same hardware architecture and performance class.
- Limited flexibility of the mainstream parallelization approaches to support distributed application scenarios—many parallel applications rely on the Single Instruction Multiple Data (SIMD) technique, which mainly targets densely built compute systems like HPC. However, the applications that are running on the truly distributed infrastructures (HPC + Cloud + remote embedded systems) have to be developed according to the MIMD approach, in order to allow different functionalities to be executed on different types of systems.

Although the mainstream simulation packages like OpenFOAM or ANSYS-CFX offer a rich development functionality which is sufficient for the implementation of the ventilation models, there are still numerous adaptations and optimizations necessary, which are very difficult to implement with the general-purpose simulation tools. For example, it would be difficult to integrate the model for diffusion and filtration processes between the airways and gas emission sources in the ventilation section. However, the major disadvantage that remains is the inability of their use in distributed, heterogeneous hardware architecture environments. Therefore, novel approaches are required for the implementation of portable, scalable, and efficient simulation software.

## 3 Microservice Architecture for Simulation Applications Development

Ventilation networks analysis is a challenging process—the simulation software developers often face problems, some of which are listed below:

- Nonlinearity of the base equation system, which causes the need of applying a special numerical method, e.g., the Finite Differences, Finite Volumes, Discontinuous Galerkin, etc.

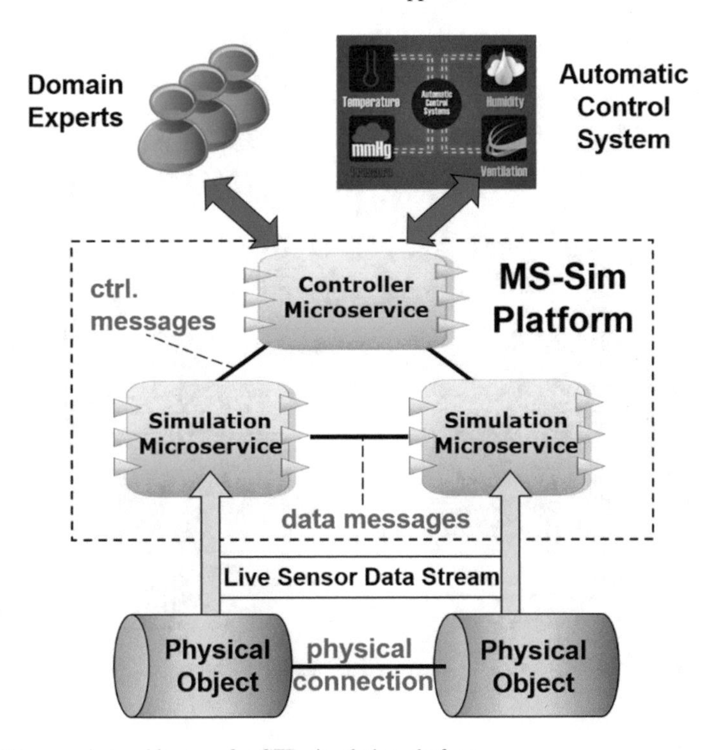


**Fig. 5** Microservice architecture for CFD simulation platform

- Complex topological organization of ventilation networks.
- Complex hierarchical structure of ventilation elements involving several levels of control and regulation.

Object-oriented modelling and service-oriented platforms have been established in the last decade as an alternative to the traditional software development approaches. The actual trend in the service-oriented development goes in the direction of microservice (MS) architectures—a concept that is initially coming from the Internet-of-Things, Cyber-Physical Systems and Cloud domains. MS are independent, isolated, portable software blocks/units, each implementing a part of a complex system, which can be decomposed according to functional, spatial or any other conditions. Each MS follows in its implementation the locality principle (as illustrated by Fig. 5)—i.e., bearing responsibility for the assigned part of the complex system, according to the decomposition strategy. In order to reflect physical or informational connections of the real object, MS can be bound together by a common data and/or control flow.

An example of a platform for execution of MS-based simulation workflows is presented in Fig. 6. The connection between all the MS in the system is provided by an external communication library, so that the MS-developers do not need to handle the data exchange explicitly—the data exchange performs asynchronously with the help of special buffers, used to flush the output data or read the input from the other

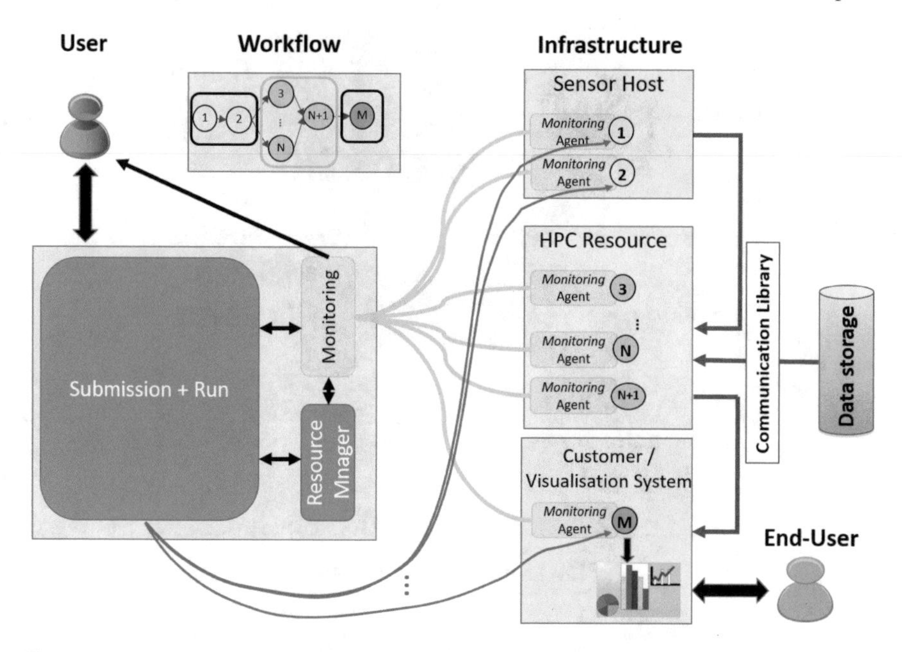


**Fig. 6** Execution platform design for microservice-based application

MS of the system, whenever required by the modelling algorithm. In case of an MPI-based implementation, every MS is executed by an independent MPI process, which is developed on the remote resource or a compute node. The MPI processes can run on heterogeneous architecture, as they could be enabled by modern implementations like OpenMPI. Each simulation MS follows a command flow, as defined by the simulation logic, which can be implemented in an event-driven way. The command flow can be steered by a dedicated "master" MS, depicted as a "Controller" in Fig. 5.

The communication between the MS happens with the help of underlying communication library, e.g., by means of point-to-point or collective MPI calls to an MPI implementation. Results storing can happen either individually by every MS (e.g., in the Paraview format) or in the collective way using a consolidating database like ElasticSearch. In fact, such functional decomposition-driven approach to the development of simulation software is not particularly new to the simulation of complex dynamic system. For example, the Matlab/Simulink modelling package provides a module-(block-)based approach to construct a model from many smaller subsystems (functional blocks or submodels). However, this approach is inefficient when dealing with big, dynamic configurations of objects with a complex and variable network topology, like the targeted ventilation systems of coalmines. The main advantages of the MS-approach for application development are the separation of the computation and communication application logic during the development process, resulting in a decrease of implementation efforts, simplified implementation of the horizontally-scalable applications (just by replicating the services), easy implementation of hierarchical relationships between the services of different functionality levels (horizontal scalability), and the possibility to deploy a MS on any resource of the heterogeneous infrastructure. In the work that is presented in the paper, a modular architecture for development of CFD simulation applications based on a library of MS-components has been implemented. A simulation application is logically organized as a modular assembly of various MS within a common architecture according to the hierarchical composition of the elements/services, as was previously depicted in Fig. 4. Moreover, using the modular approach, the MS-based simulation application can transparently perform experimentation with different approximation schemes, discretization approaches, numerical solution methods, results validation techniques, etc.

## 4 Implementation of CFD Ventilation Study with Microservices Architecture

The above-presented microservices approach was used to implement a simple CFD study for a ventilation section consisting of 2 airways ($Q_{FW}$, $Q_{VS}$), a coal mining area ($Q_S$), and a goaf (q) with a gas emission source ($q_m$), according to the topology depicted in Fig. 7. The airflow was enforced by a single mine fan, connected to the first airway ($Q_{FW}$).

Each element of the section is described by the base model (1). However, since the system (1) includes partial differential equations, the models development had to start from a lower granularity level—approximation elements of the numerical method of spatial discretization. In our case, the finite differences method was chosen for the reasons of simplicity), that resulted in a set of k normal differential equations for approximation units in the general form (2):

$$\begin{cases} \frac{dQ_k}{dt} = \frac{F}{\rho} \cdot \frac{P_k - P_{k+1}}{\Delta y} - \frac{F}{\rho} r Q_k^2 - \frac{F}{\rho} r(t) \cdot Q_k^2 \\ \frac{dP_{k+1}}{dt} = \frac{\rho a^2}{F} \cdot \frac{Q_k - Q_{k+1}}{\Delta y} - \frac{\rho a^2}{F} q_k, \end{cases} \quad (2)$$

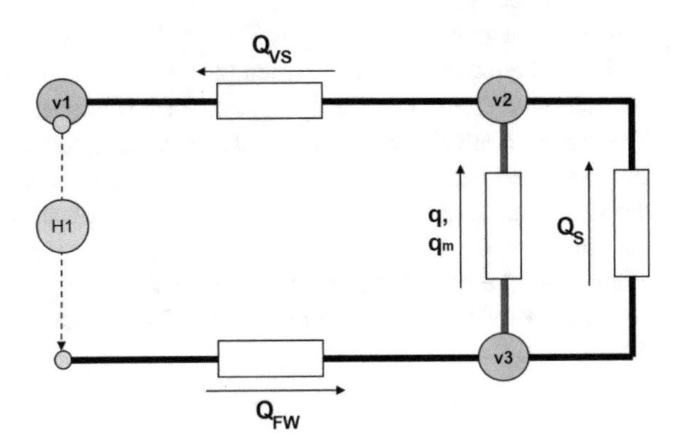


**Fig. 7** Test section structure

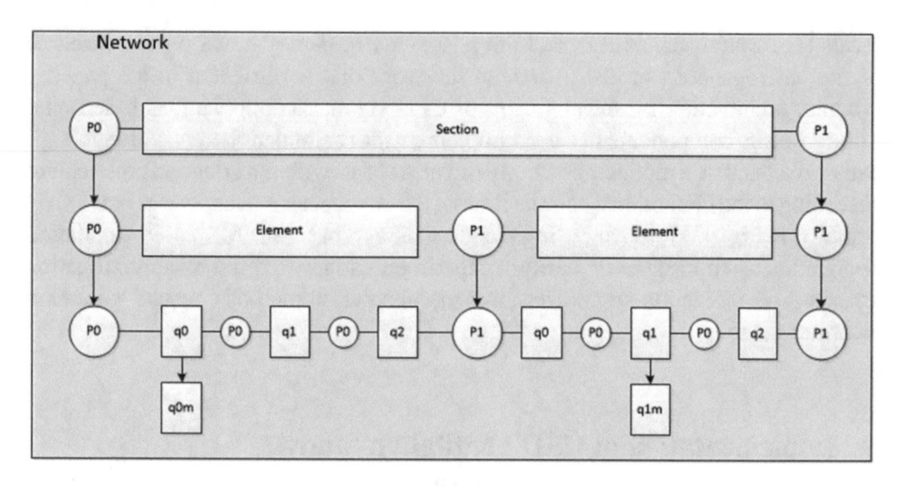


**Fig. 8** Hierarchical composition of modelling services

where the first equation represents the fluid transport in an (spatial) approximation element, and the second—conservation in approximation nodes, as elaborated in the previous publication [4]. These models form the bottom level of the ventilation models hierarchy (cf. Fig. 4) and used for creation of the models of all upper hierarchical levels (airways/elements, sections, network), as shown in Fig. 8. Connections between the models (or precisely—their corresponding services) corresponded to the boundary conditions of the equations (2).

The models were implemented with Open MPI as the underlying deployment and communication library. Three different scenarios were tested (A, B, C), as shown in Fig. 9—increase of fan pressure P (scenario A), adjustment of global regulator r (scenario B), and drop of fan pressure P (scenario C). The results are shown for flow in the outbound element ($Q_{VS}$) and goaf (q), as well as marsh gas concentration C, which was defined as a ratio $q_m/q$. All experiments (A, B, and C) showed results as expected to be by the physical experiments (e.g., performed by Svjatnyj in [8]). The service-based application workflow was steered by a dedicated supervising service—a manager, which performs the following functions:

The dynamic approach, in which the services act as independent interactive components which are continuously running on the dedicated hardware and can be steered by a remote controller according to the specific application logic is particularly interesting for real-time control scenarios. In such scenarios, the services can incorporate the sensor data, make prediction for the future situation development, and instruct the control system about the probability of any potential risks appearance. On the other hand, the models can be optimized by adapting their parameters to best fit the actual mode of the controlled complex dynamic system.

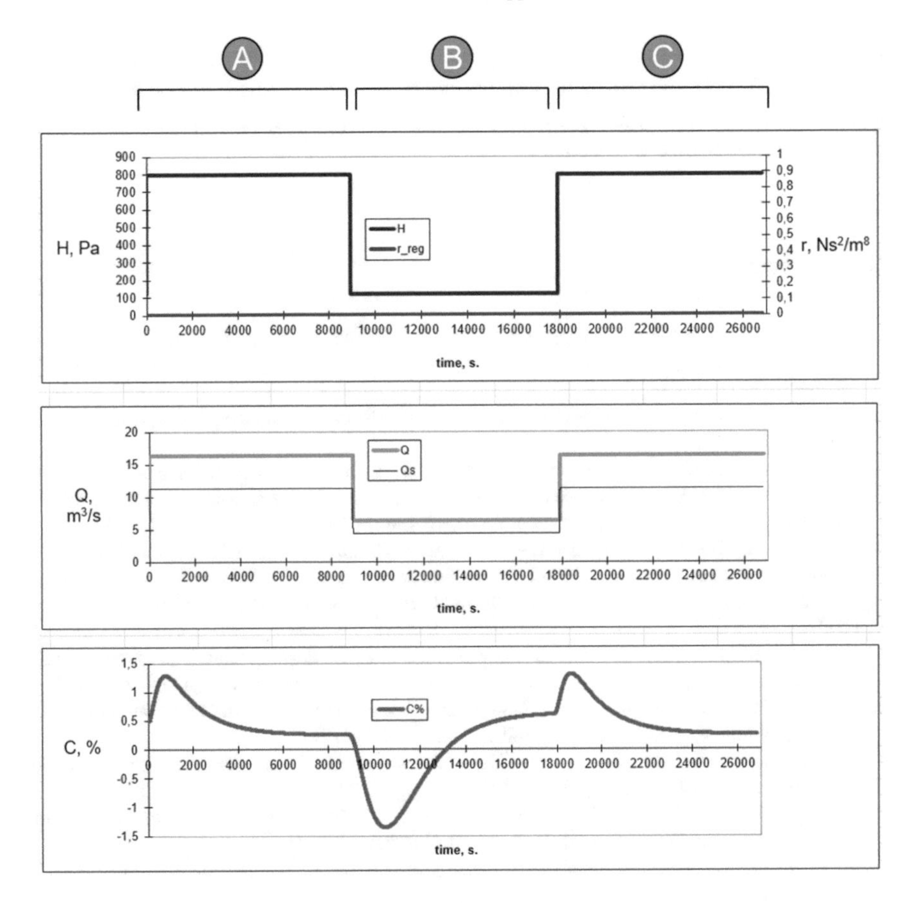


**Fig. 9** Modelling results

## 5 Conclusion

The simulation technology is facing the challenges of application for new real-time scenarios that require a high flexibility of modelling tools in terms of the broader usage of the available infrastructure (data acquisition, storage, and processing devices). The rapid development of sensor networks has made possible a number of new innovative scenarios, for which the monolithic design of the existing simulation tools and workflow solutions on their top might be of a big obstacle. Service-oriented platforms offer a promising vision of the future development of simulation tools by offering benefits of on-demand distribution and parallelization, which might be well supported the underlying management platforms. Microservices are the technology which can if not fully replace the workflow-based scenarios but have the potential to support and bring them to the principally new level of usability. The effort that was done on implementation of the ventilation scenario has revealed a high potential of microservices architectures in geoscience and other domains of science and technology. The further research will concentrate, among others, on elaboration of functional composition strategies to development of complex, assembled service for hierarchically-organized systems.

## References

1. Andruleit, H., et al.: BGR Energie Study 2018—Data and Development Trends of German and Global Energy Supplement. Bundesanstalt für Geowissenschaften und Rohstoffe (BGR), Hannover (2019)
2. Broy, M.: Cyber-Physical Systems. Innovation through Software-Intensive Embedded Systems. Springer (2010)
3. Chang, P., et al.: Cyberinfrastructure requirements to enhance multi-messenger astrophysics. In: Proceedings of the Astro2020: Decadal Survey on Astronomy and Astrophysics. Science White Papers, no. 436. Bull. Am. Astron. Soc. **51**(3), id. 436 (2018)
4. Cheptsov, A.: The system organization and basic algorithms of the simulation and servicing center for the coal industry. In: IEEE Proceeding International Conference Modern Problems of Radio Engineering,Telecommunications and Computer Science TCSET'2007, pp. 205–207 (2007)
5. Cheptsov, A.: From static to dynamic: a new methodology for development of simulation applications. In: Advances in Intelligent Systems: Reviews, Book Series, vol. 1, pp. 69–88 (2017)
6. Clausen, E.: Mine ventilation in the 21st century—development towards adaptive ventilation systems. Mining Rep. Glückauf **153**(4/2017), 326–332 (2017)
7. Stewart, C., Aminossadati, S., Kizil, M.: Use of live sensor data in transient simulations of mine ventilation models. Mining Rep. **153**(4/2017), 356–363 (2017)
8. Svjatnyj, V.: Simulation of Air Dynamic Processes and Development of Control System for Underground Mine Ventilation. Ph.D. thesis (in Russian) (1985)

# Containerization and Orchestration on HPC Systems

**Naweiluo Zhou**

**Abstract** Containerization demonstrates its efficiency in application deployment in Cloud clusters. HPC systems start to adopt containers, as containers can encapsulate complex programs with their dependencies in isolated environments making applications more portable. Nevertheless, conventional HPC workload managers lack micro-service support and deeply-integrated container management, as opposed to container orchestrators. We enable the synergy of Cloud and HPC clusters. We propose the preliminary design of a feedback control scheduler that performs efficient container scheduling meanwhile taking advantage of the scheduling policies of the container orchestrator (Kubernetes) and the HPC workload manager (TORQUE).

## 1 Introduction

Containerization can ensure application compatibility with environments by encapsulating programs with their dependencies [22], which enables users to move and deploy programs easily among Cloud clusters. Containerization is a virtualization technology [37]. A container only shares the host OS in lieu of starting a holistically simulated OS on top of the host OS as in a Virtual Machine (VM). Consequently, containers are more lightweight than VMs. Often containers are dedicated to run micro-services and one container mostly hosts one application. Nevertheless, containerized applications can become complex, e.g. thousands of separate containers may be required in production. Production can benefit from container orchestrators that can provide efficient environment provisioning and auto-scaling.

Applications on Cloud clusters are becoming more data intensive and computation demanding, especially with the development of Artificial Intelligence (AI) applications. High Performance Computing (HPC) systems are traditionally applied to perform large-scale financial and engineering simulation which demand low-latency and high-throughput. HPC systems are typically equipped with workload managers. A *workload manager* is composed of a *resource manager* and a *job scheduler*. A

N. Zhou (✉)
High Performance Computing Centre Stuttgart, Stuttgart, Germany
e-mail: naweiluo.zhou@hlrs.de



M. M. Resch et al. (eds.), *Sustained Simulation Performance 2019 and 2020*,
https://doi.org/10.1007/978-3-030-68049-7_10

resource manager [16] allocates resources (e.g. CPU and memory), schedules jobs and guarantees no interference from other user processes. A job scheduler determines the job priorities, enforces resource limits and dispatches jobs to available nodes [23]. HPC workload managers lack micro-service supports and deeply-integrated container management capabilities in which container orchestrators manifest the efficiency.

Due to the high energy cost of HPC clusters, e.g. a 100-MW data center that only wastes 1% of its computing cycles can nullify all the energy-saving measures of a small city [36], high resource utilization is usually the major scheduling concern. We adopt the architecture implemented in our previous work [50], which enables container scheduling from the container orchestrator (Kubernetes) to the HPC workload manager (TORQUE). Based on this architecture, we present our preliminary design of a Kubernetes scheduler that makes efficient resource scheduling via a feedback control loop meanwhile taking advantage of both the scheduling policies on Kubernetes and TORQUE.

The remainder of the paper is organized as follows. Firstly, we present the related background on the key techniques and technologies in Sect. 2. Next, Sect. 3 reviews the related work. Followed, we describe the scheduler design and briefly present the system architecture in Sect. 4. Lastly, Sect. 5 concludes this paper and proposes future work.

## 2 Background

This section gives the core background techniques and technologies on containerization, resource management and scheduling for Cloud and HPC clusters.

### *2.1 Containerization*

A container is an OS-level virtualization technique that provides execution environment separation for applications. A container encapsulates programs with their corresponding libraries, data, configuration files, etc. [22] in an isolated environment, which ensures package compatibility, thus enables users to deploy programs easily on clusters. Figure 1 presents the architecture difference of VMs and containers. A Traditional VM loads an entire guest OS (the simulated OS) into memory, which can occupy gigabytes of storage space and require a significant fraction of system resources to run. *Per contra*, a container can utilize the dependencies on its host OS. To boot a new container, the host merely needs to start new processes that are isolated from its own, which makes start-up time of a container similar to that of a native application [1, 7, 12, 46]. Apart from their portability, containers can guarantee reproducibility, i.e. once a workflow has been defined and stored in the container, its included working environment remains unchanged regardless of its execution occur-

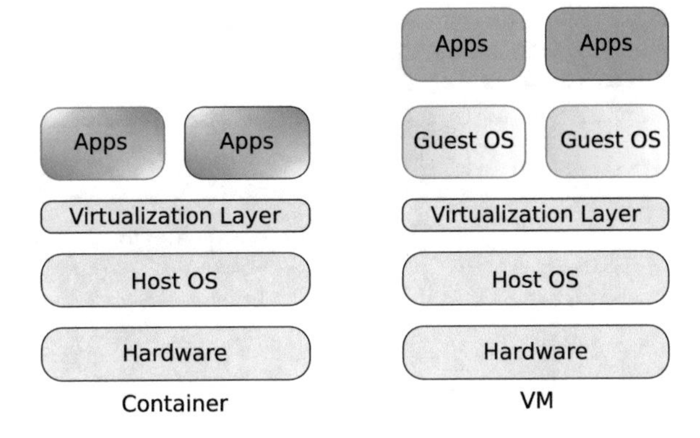


**Fig. 1** Structure comparison of VMs and containers. On the VM side, the virtualization layer often appears to be Hypervisor while on the container side it is the container runtime

rences. Containers can also run inside VMs as this is the case on most of the Cloud clusters [6].

There are multiple technologies that realise the concept of containers, e.g. Docker [28], Singularity [24], Shifter [10], Charlie Cloud [35], Linux LXC [39] and Rkt Core OS [27]. The Singularity container runtime was specifically designed in the first place for HPC systems. It manifests the following merits:

- Run with user privileges and need no daemon process. Acquisition of root permission is only necessary when users want to build or rebuild Singularity images, which can be performed on their own working computers. Unprivileged users can also build images from a definition file by Singularity "fake root" with a few restrictions, nevertheless, some methods requiring to create block devices (e.g. `/dev/null`) may not always work correctly;
- Seamless integration with HPC. Singularity natively supports GPU, Message Passing Interface (MPI) [11] and InfiniBand. Additional network configuration is not needed. Generating an HPC environment inside Singularity is as easy as installing software packages on a Unix-like system;
- Portable via a single image file.

Singularity has, therefore, become the *ipso facto* standard container runtime for HPC systems.

## 2.2 Workload Managers on HPC

Two main-stream workload managers are PBS [41] and Slurm [19]. This section only gives a description of PBS. The principle of Slurm is similar.

PBS stands for Portable Batch System which includes three versions: TORQUE, PBS Pro and OpenPBS. We herein briefly introduce TORQUE as the three share similar mechanisms.

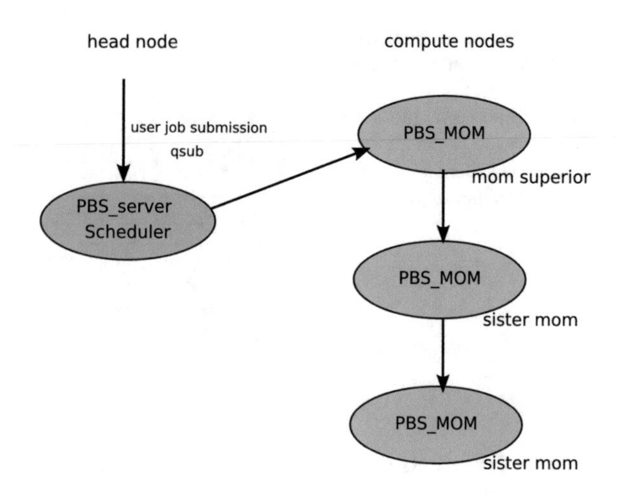


**Fig. 2** Torque structure. A head node controls the rest of the compute nodes

The structure of TORQUE is composed of a head node and many compute nodes as illustrated in Fig. 2. The head node is the master node that controls the entire TORQUE system. A *pbs_server* and a job scheduler daemon are located on the head node. The batch job is submitted to the head node (in some cases, especially on an HPC production system, jobs are first submitted to a login node and then transferred to the head node). A node list is maintained on the head node which specifies the list of compute nodes in the cluster. We herein briefly describe the procedure of job submission on TORQUE as follows:

1. the job is submitted to the head node by TORQUE command `qsub`. The job can be written in a PBS job script or simply include in the `qsub` command. A job ID will be generated and returned to the user as standard output of `qsub` if the job is accepted.
2. The job record, which incorporates job ID and job attributes, is generated and passed to *pbs_server*.
3. *pbs_server* transfers the job record to the job scheduler (e.g. Moab) daemon. The job scheduler daemon adds the job into a job queue and applies a scheduling algorithm to it (e.g. FIFO) which determines the job priority and its resource assignment.
4. When the scheduler finds the list of nodes for the job, it returns the job information to *pbs_server*. The first node on this list becomes the Mother superior and the rest are called sister MOMs or sister nodes. *pbs_server* allocates the resources and passes the job control as well as execution information to *pbs_mom* daemon installed on the mom superior node which will be instructed to launch the job on the assigned compute nodes.
5. The *pbs_mom* daemons on the compute nodes manage execution of jobs and monitors resource usage. *pbs_mom* will capture all the outputs and direct them to *stdout* and *stderr* files of the job and copy them to the designated location when

the job completes successfully. The job status (completed or terminated) will be passed to *pbs_server* by *pbs_mom*. The job information will be updated.

Before and after a job execution, TORQUE executes *prologue* and *epilogue* scripts to prepare systems and perform node health check, append text to output and error log files, clean up system, etc.

In TORQUE, nodes are partitioned into different groups called *queues*. In each queue, resource limits are set, such as walltime and job size. This feature can be useful for job scheduling in a large HPC cluster where nodes are heterogeneous or certain nodes are reserved for special users. One node can be included in several queues.

TORQUE is commonly integrated with a sophisticated scheduler, such as Maudi. The Maudi Scheduler [18] is an open source job scheduler that provides advanced features such as dynamic job prioritization, configurable parameters, extensive fair share capabilities and backfill scheduling. Maudi functions in an iterative manner. It starts a new iteration when one of the following conditions is met: (1) a job or resource state alters, (2) a reservation boundary event occurs (3) an external command to resume scheduling is issued (4) a configuration timer expires. In each iteration, Maudi follows the steps as below [34]:

1. Obtain resource information from TORQUE;
2. Fetch workload information from TORQUE;
3. Update statistics;
4. Refresh reservations;
5. Select jobs eligible priority scheduling;
6. Prioritise eligible jobs;
7. Schedule the jobs in priority order and create reservations;
8. Backfill jobs.

Backfilling scheduling [29] allows jobs to take the reserved job slots provided that the start of the other jobs having reserved the resources will not be delayed, thus allowing large parallel jobs to execute without under utilizing resources.

## 2.3 *Resource Managers on Big Data Clusters*

It requires an effective scheduling mechanism to deal with the massive amount of data in a distributed cluster such as big-data cluster on the Cloud. Many frameworks [14, 15, 21, 40, 42, 43, 47, 48] have been widely used in academia or industry, we give introduction to two frameworks that are among the most popular ones.

### 2.3.1 Kubernetes

Thousands of separate containers may be required in production, which can benefit from container orchestrators that can offer [3, 4, 14]:

- Resource limit control. Kubernetes reserves a specific amount of CPU and memory for a container, which restrains interference among containers and provides information for scheduling decision;
- Scheduling. Determine the policies on node placement optimization for containers;
- Load balancing. Distribute the workload among container instances;
- Heath check. Verify if a faulty container needs to be destroyed or replaced;
- Fault tolerance. Create containers automatically if applications or nodes fail, hence it allows to maintain a desired number of containers;
- Auto-scaling. Add and remove containers automatically.

Kubernetes is among the most advanced open-source container orchestrators, which has a rapidly growing community and ecosystem with numerous platforms being developed upon it. The architecture of Kubernetes is composed of a master node and many worker nodes. Kubernetes runs containers inside *pods* that are scheduled to the master or worker nodes. A *pod* can encapsulate one or multiple containers. Kubernetes provides its services by *deployments* that are created by submission of *yaml* files. Inside the *yaml* file, users can specify services and computation to perform. A user *deployment* can be performed either on the master node or the worker nodes.

Kubernetes is based on a highly modular architecture which abstracts the underlying infrastructure and allows internal customization, such as deployment of different software defined networks or storage solutions. It also supports various Big-Data frameworks, such as Hadoop MapReduce [31], Spark [48] and Kafka [30]. It includes a powerful set of tools to control the life cycle of applications, e.g. parameterized redeployment in case of failures, state management, etc. Furthermore, Kubernetes supports software defined infrastructures[1] and resource disaggregation [9] by leveraging container-based deployments and particular drivers (e.g. Container Runtime Interface driver, Container Storage Interface driver and Container Network Interface driver) based on standardized interfaces. These interfaces enable the definition of abstractions for fine-grain control of computation, states and communication in multi-tenant Cloud environment along with optimal usage of the underlying hardware resources.

Kubernetes incorporates an advanced scheduling system which even allows to specify a different scheduler for each job. The scheduling system makes its decisions based on three steps:

1. Node filtering. The scheduler locates the node/nodes that fit the workload, e.g. a pod specifies node affinity constraints making only certain nodes meet its requirements, or some nodes do not contain enough CPU resources for the pod request. Normally the scheduler does not go through the entire node list, rather it selects the one/ones detected first.

---

[1]Software-defined infrastructure (SDI) is the definition of technical computing infrastructure entirely under the control of software with no operator or human intervention. It operates independent of any hardware-specific dependencies and is programmatically extensible.

2. Node priority calculation. The scheduler calculates a score for each node, and the highest scoring node will run that pod.
3. Actual scheduling operations.

The traditional HPC workload manager lacks its efficiencies in container scheduling and management, and often do not provide integrated support for environment (i.e. infrastructure, configuration and dependency) provisioning. Moreover, HPC applications are hardware specific, and are often specifically optimized for the nodes. This is not the case for containerized applications. Considering that performance is *sine qua non* for HPC applications, it poses the crucial question for massive usage of containerized applications on HPC clusters [45]. Nevertheless, some works have denoted near-native performance achieved by containers under certain situations [17, 27, 33, 49].

### 2.3.2 Apache Mesos

Apache Mesos [15, 21] is a cluster manager that provides efficient resource isolation and sharing across distributed applications or frameworks. Mesos is consisted of a master process that manages slave daemons running on each cluster node and frameworks that run tasks on the slave nodes. It takes control over scheduling to frameworks through a *resource offer* that encapsulates resources which a framework can allocate on a cluster node to run tasks. The *resource offer*, which resides on the master node, is a list of free resources on multiple slave nodes. The master node determines the amount of resources that can be given to each framework according to policies of each organization, such as fair sharing. Each framework on the slave nodes incorporates two components, i.e. a scheduler registered to the master node to receive resources and an executor process to run tasks of the frameworks. When a framework accepts the offered resources, it passes Mesos a description of the tasks that launches on it. Figure 3 illustrates an architecture of Mesos that connects with two different frameworks: Hadoop [43] and MPI.

## 3 Related Work

A few studies [13, 20, 22] have been carried out on container orchestration for HPC systems. Some works [25, 32, 50] have been performed on the general issues of bridging the gap between conventional HPC and service-oriented infrastructures. Literature has shown numerous works [2, 3, 8, 26] on container orchestration for Cloud clusters, though this is out of the scope herein.

This work is based on the architecture that we proposed in [50]. In that study, we described a plugin named Torque-Operator which enables container scheduling from Kubernetes to TORQUE. The proposed architecture converges HPC and Cloud systems which are located in two different network domains. Nevertheless, the study

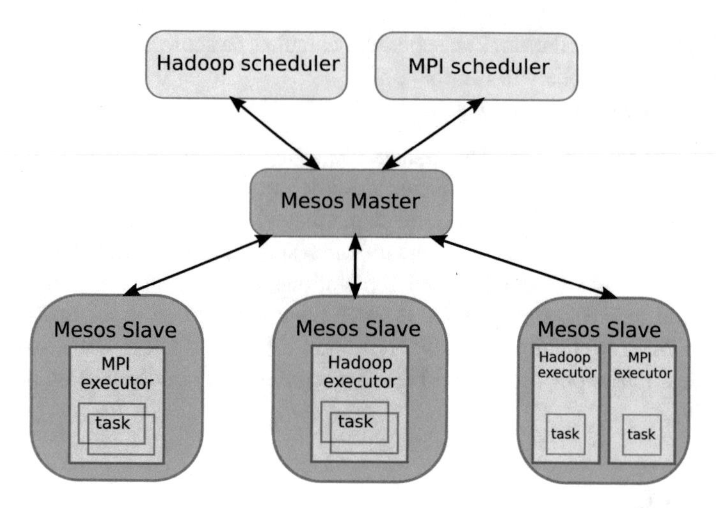


**Fig. 3** A Mesos architecture [15] with two running frameworks: Hadoop and MPI

has not given insights into container scheduling policies that will be presented later herein.

Liu et al. [25] showed how to dynamically migrate computing resources between HPC and OpenStack [38] clusters based on demand. At a higher level, IBM has demonstrated the ability to run Kubernetes *pods* on Spectrum LSF (an HPC workload manager).

Piras et al. [32] expanded Kubernetes clusters onto HPC clusters through Grid Engine (an HPC workload manager). Its submission is performed by the PBS job to launch Kubernetes jobs. Therefore, HPC nodes are added to Kubernetes clusters by installing Kubernetes core components (i.e. *kubeadm*, *Kubelet*) and Docker container runtime on the HPC nodes. On HPC systems, especially HPC production systems, it is not always straightforward to add new software packages, as it can cause security concerns and alter working environment for current users.

Khan et al. [22] proposed to install Mesos and Marathon[2] on HPC clusters for resource management and container orchestration. The HPC jobs are containerized. This orchestration system can obtain the appropriate resources satisfying the needs of requested services within defined Quality-of-Service (QoS) parameters, which is stated to be self-organized and self-managed meaning that users do not need to specifically request resources reservation. Nevertheless, this study has not shown insight into novel strategies of container orchestration for HPC systems.

Julian et al. [20] presented their prototype for container orchestration in an HPC environment. A PBS based HPC cluster that can automatically scale up and down as load demands by launching Docker containers using Moab scheduler. Three containers serve as the front-end system, scheduler (it runs workload manager inside) and compute node (launches *pbs_mom* daemon). More compute node containers are

[2]Marathon container orchestrator: https://mesosphere.g.ithub.io/marathon/.

scheduled when there are no sufficient number of physical nodes. Unused containers are destroyed via external Python scripts when jobs complete. Similarly, an early study [13] described two models that can orchestrate Docker containers using HPC resource manager. The former model launches a container to mimic one compute node which holds all the processes assigned to, whilst the latter boots a container per process.

Wrede et al. [44] performed their experiments on HPC clusters using Docker Swarm [40] as the container orchestrator for automatic node scaling and using C++ algorithmic skeleton library Muesli [5] for load balance. Nevertheless, its proposed working environment is targeted for Cloud clusters. Usage of Docker can not be easily extended to HPC infrastructures, as Docker can raise security concerns.

## 4 Container Orchestration on HPC Systems

We connect a Cloud cluster (VM nodes managed by Kubernetes) with an HPC cluster (bare-metal nodes controlled by TORQUE). The computation-intensive or data-intensive jobs on the Cloud cluster can be scheduled to the HPC cluster for better performance. This is achieved by the architecture proposed in our former work [50], which will be briefly presented in Sect. 4.1.

On Kubernetes, containers are scheduled in the unit of *pods*. Its scheduler determines the best node/nodes for the *pods* to execute on. We implement a feedback control scheduler that takes the default scheduler of Kubernetes as its base. The job scheduling policy stays unaltered on the HPC side. The feedback control scheduler can enhance efficiency of container scheduling from the Cloud cluster to the HPC cluster.

### *4.1 Architecture Description*

As illustrated in Fig. 4, the architecture includes a Cloud cluster consisting of 4 VM nodes and an HPC cluster consisting of 5 HPC nodes grouped into two queues. Nevertheless, the number of queues is not limited to the amount as shown. What matters in this architecture is the login node as highlighted in the red dashed line, which provides a unified job submission interface and enables container scheduling from Kubernetes to TORQUE. The login node resides in both clusters. It is a Kubernetes worker node and meanwhile a login node of TORQUE. The login node only enables job submission to the TORQUE head node and the jobs from HPC cluster will not be scheduled to this node.

The job is written in a *yaml* script supported by Kubernetes. The TORQUE job is encapsulated inside the script as shown in Fig. 5 from Line7 to Line 14. More specifically, the job requests 30 min walltime and one compute node. It runs a Singularity image called *lolcow.sif* located in the `home` directory, which is pulled

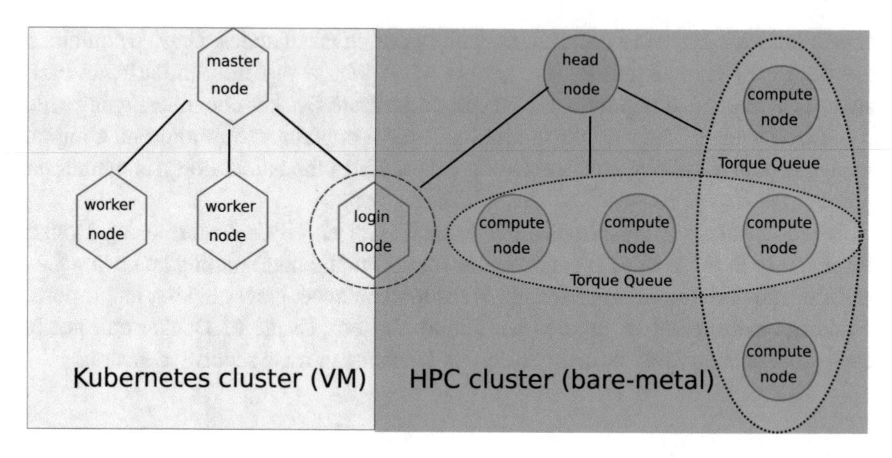


**Fig. 4** Architecture of the platform. The HPC cluster contains two queues

```
1  apiVersion: wlm.sylabs.io/v1alpha1
2  kind: TorqueJob
3  metadata:
4    name: cow
5  spec:
6    batch: |
7      #!/bin/sh
8      #PBS -l walltime=00:30:00
9      #PBS -l nodes=1
10     #PBS -e $HOME/cow.err
11     #PBS -o $HOME/cow.out
12     export PATH=$PATH:/usr/local/bin
13     singularity pull -U library://sylabsed/example/lolcow
14     singularity run lolcow_latest.sif
15   results:
16     from: $HOME/cow.out
17     mount:
18       name: data
19       hostPath:
20         path: $HOME
21         type: DirectoryOrCreate
```

**Fig. 5** An example of the Kubernetes *yaml* script. The script encloses a PBS script. The jobs is submitted as a normal *yaml* job

from the Syslab registry. Its results and error messages will be written in *cow.out* and *cow.err*, respectively. Line 12 sets the necessary `PATH` environment variable to find the Singularity executable located in `/usr/local/bin`. The *yaml* script is accepted by Kubernetes cluster and the TORQUE script part is abstracted and re-submitted to TORQUE (refer to [50] for more details).

## 4.2 The Feedback Control Scheduler

On Kubernetes, we add a new scheduler that will be exclusively in charge of the job scheduling from Kubernetes to TORQUE. The rest of the jobs on the Kubernetes cluster still obey the default scheduling policies. The default scheduler on Kubernetes is not aware of the resource availability of each HPC queue, e.g. when there are not enough free nodes on the HPC cluster, Kubenetes scheduler can continue to assign the new jobs to HPC, therefore, the submissions will be listed in the TORQUE waiting queue.

The feedback control scheduler is implemented on Kubernetes. It takes the HPC node usage as its inputs, follows the Round-Robin algorithm, and outputs the decision on which TORQUE queue to submit the job. When the available nodes in one TORQUE queue does not satisfy the request, the jobs are designated to a second queue. The decision procedure is given in Fig. 6. In the Round-Robin algorithm, the first-come job is assigned to to the first queue and the next job will be scheduled to the second queue, etc. A counter is incremented by one when the scheduler completed one round of all the queues. When the counter reaches a certain value, a node status inquiry will be issued to TORQUE head node in order to avoid frequent node status inquiries that can interrupt job execution on the HPC cluster. The initial value of the counter threshold is set to be one. It decrements by one when the scheduled queue differs from the Round-Robin decision based on queue status inquiry and it increments by one vice versa. Its maximum value is equivalent to the total amount of HPC compute nodes connected with the Cloud cluster and the minimum is one.

The job is labeled with `virtual-kubelet` which can guarantee its transfer to the HPC cluster via the *virtual node* with the same *node label* `virtual-kubelet`. One virtual node corresponds to one TORQUE queue and each virtual node contains its corresponding queue information, e.g. the total amount of nodes in the queue (refer to [50] for further details on the virtual node). Without the feedback control scheduler, scheduling decision depends on the fixed resource limits labeled on the virtual node rather than the just-in-time node availability.

The scheduler is implemented in Golang, however, it can be written in any programming languages. The scheduler program is encapsulated in a Singularity

```
1   counter_threshold =1
2   while (true)
3       counter=1;
4          //when less than the total amount of the nodes available on TORQUE
5       while (counter<= counter_threshold)
6                  Scheduling_policy=Roud-Robin;
7                  counter++;
8       PBS status inquiry;
9       if (scheduling decision alters && counter_threshold >1)
10                counter_threshold--;
11         else if (scheduling deision remains && counter_threshold <=num_of_nodes)
12                counter_threshold++;
```

**Fig. 6** The feedback control scheduling algorithm. Written in pseudo C code

image. A standard Kubernetes *deployment* is created to register the scheduler on the Kubernetes cluster. To obtain the TORQUE queue status, we issue a PBS command `pbsnodes` on the login node.

### 4.3 Discussion

The scheduler proposed herein can be suitable for a small HPC cluster where it encompasses several queues and each queue contains a small number of nodes. In an HPC production system, a queue can incorporate hundreds of nodes and the nodes can be heavily used by the workloads, thus making our scheduler less interesting. In this case, it would be more practical to leave the HPC workload manager to dictate the right queue for execution. The HPC cluster should be considered as one unit that corresponds to one *virtual node* in preference to each virtual node correlating to one TORQUE queue.

## 5 Conclusion and Future Work

We bridge the HPC cluster with the Cloud cluster. The data intensive or computation demanding applications can be scheduled from the Cloud cluster to the HPC Cluster that can significantly accelerate execution. The TORQUE and Kubernetes jobs could be submitted via the same interface on the Kubernetes. Efficient resource utilization is a major scheduling concern on HPC clusters. We proposed a feedback control scheduler that adopts the default Kubernetes scheduler as its base and makes scheduling decision by taking into node availability on HPC queues into consideration.

The future work will be carried out on scheduling efficiency evaluation. Its benchmark performance can be compared with the default Kubernetes scheduler. Another direction can be performed to extend the TORQUE scheduler to support sophisticated container orchestration.

**Acknowledgements** This project has received funding from the European Union's Horizon 2020 research and innovation programme under grant agreement NO. 825355.

## References

1. Bernstein, D.: Containers and cloud: from LXC to Docker to Kubernetes. IEEE Cloud Comput. **1**(3), 81–84 (2014)
2. Buyya, R., Srirama, S.N.: A lightweight container middleware for edge cloud architectures, pp. 145–170. Wiley (2019)
3. Casalicchio, E.: Container orchestration: a survey. In: Puliafito, A., Trivedi, K.S. (eds.) Systems Modeling: Methodologies and Tools, pp. 221–235. Springer International Publishing, Cham (2019)

4. Casalicchio, E., Iannucci, S.: The state-of-the-art in container technologies: application, orchestration and security. Concurr. Comput. Pract. Exp. e5668 (2020)
5. Ciechanowicz, P., Poldner, M., Kuchen, H.: The M*ü*nster skeleton library Muesli: a comprehensive overview
6. Containers on virtual machines or bare metals? Tech. Rep., VMWare, Dec 2018
7. Felter, W., Ferreira, A., Rajamony, R., Rubio, J.: An updated performance comparison of virtual machines and Linux containers. In: 2015 IEEE International Symposium on Performance Analysis of Systems and Software (ISPASS), pp. 171–172 (2015)
8. Fernandez, G.P., Brito, A.: Secure container orchestration in the cloud: policies and implementation. In: Proceedings of the 34th ACM/SIGAPP Symposium on Applied Computing, SAC 19, New York, NY, USA, pp. 138–145. Association for Computing Machinery (2019)
9. Gao, P.X., Narayan, A., Karandikar, S., Carreira, J., Han, S., Agarwal, R., Ratnasamy, S., Shenker, S.: Network requirements for resource disaggregation. In: Proceedings of the 12th USENIX Conference on Operating Systems Design and Implementation, USA, OSDI 16, pp. 249–264. USENIX Association (2016)
10. Gerhardt, L., Bhimji, W., Canon, S., Fasel, M., Jacobsen, D., Mustafa, M., Porter, J., Tsulaia, V.: Shifter: containers for HPC. J. Phys. Conf. Ser. **898**, 082021 (2017)
11. Gropp, W., Lusk, E., Skjellum, A.: Using MPI: Portable Parallel Programming with the Message-passing Interface. MIT Press, Cambridge, MA, USA (1994)
12. Hale, J.S., Li, L., Richardson, C.N., Wells, G.N.: Containers for portable, productive, and performant scientific computing. Comput. Sci. Eng. **19**(6), 40–50 (2017)
13. Higgins, J., Holmes, V., Venters, C.: Orchestrating docker containers in the HPC environment. In: Kunkel, J.M., Ludwig, T. (eds.) High Performance Computing, pp. 506–513. Springer International Publishing, Cham (2015)
14. Hightower, K., Burns, B., Beda, J.: Kubernetes: Up and Running Dive into the Future of Infrastructure, 1st edn. O'Reilly Media, Inc. (2017)
15. Hindman, B., Konwinski, A., Zaharia, M., Ghodsi, A., Joseph, A.D., Katz, R., Shenker, S., Stoica, I.: Mesos: a platform for fine-grained resource sharing in the data center. In: Proceedings of the 8th USENIX Conference on Networked Systems Design and Implementation, USA, NSDI 11, pp. 295–308. USENIX Association (2011)
16. Hovestadt, M., Kao, O., Keller, A., Streit, A.: Scheduling in hpc resource management systems: queuing vs. planning. In: Feitelson, D., Rudolph, L., Schwiegelshohn, U. (eds.) Job Scheduling Strategies for Parallel Processing. Springer, Berlin, Heidelberg (2003)
17. Hu, G., Zhang, Y., and Chen, W. Exploring the performance of singularity for high performance computing scenarios. In: 2019 IEEE 21st International Conference on High Performance Computing and Communications; IEEE 17th International Conference on Smart City; IEEE 5th International Conference on Data Science and Systems (HPCC/SmartCity/DSS), Aug 2019, pp. 2587–2593
18. Jackson, D., Snell, Q., Clement, M.: Core Algorithms of the Maui scheduler. In: Feitelson, D.G., Rudolph, L. (eds.) Job Scheduling Strategies for Parallel Processing, pp. 87–102. Springer, Berlin, Heidelberg (2001)
19. Jette, M.A., Yoo, A.B., Grondona, M.: SLURM: simple linux utility for resource management. In: Proceedings of Job Scheduling Strategies for Parallel Processing (JSSPP) 2003. Lecture Notes in Computer Science, pp. 44–60. Springer (2002)
20. Julian, S., Shuey, M., Cook, S.: Containers in research: initial experiences with lightweight infrastructure. In: Proceedings of the XSEDE16 Conference on Diversity, Big Data, and Science at Scale, New York, NY, USA, 2016, XSEDE16. Association for Computing Machinery (2016)
21. Kakadia, D.: Apache Mesos Essentials. Packt Publishing (2015)
22. Khan, M., Becker, T., Kuppuudaiyar, P., Elster, A.C.: Container-based virtualization for heterogeneous HPC clouds: insights from the EU H2020 cloudlightning project. In: 2018 IEEE International Conference on Cloud Engineering (IC2E), Apr 2018, pp. 392–397 (2018)
23. Klusáček, D., Chlumský, V., Rudová, H.: Planning and optimization in torque resource manager. In: Proceedings of the 24th International Symposium on High-Performance Parallel and Distributed Computing, New York, NY, USA. Association for Computing Machinery (2015)

24. Kurtzer, G.M., Sochat, V.V., Bauer, M.: Singularity: scientific containers for mobility of compute. PloS one (2017)
25. Liu, F., Keahey, K., Riteau, P., Weissman, J.: Dynamically negotiating capacity between on-demand and batch clusters. In: Proceedings of the International Conference for High Performance Computing, Networking, Storage, and Analysis, SC 18. IEEE Press (2018)
26. Maenhaut, P.-J., Volckaert, B., Ongenae, V., De Turck, F.: Resource management in a containerized cloud: status and challenges. J. Netw. Syst, Manag. (2019)
27. Martin, J.P., Kandasamy, A., Chandrasekaran, K.: Exploring the support for high performance applications in the container runtime environment. Hum. Centric Comput. Inf. Sci. **8**, 1 (2018)
28. Merkel, D.: Docker: lightweight linux containers for consistent development and deployment. Linux J. **2014**, 239 (2014)
29. Mu'alem, A.W., Feitelson, D.G.: Utilization, predictability, workloads, and user runtime estimates in scheduling the IBM SP2 with backfilling. IEEE Trans. Parallel Distrib. Syst. **12**(6), 529–543 (2001)
30. Narkhede, N., Shapira, G., Palino, T.: Kafka: The Definitive Guide Real-Time Data and Stream Processing at Scale, 1st edn. O'Reilly Media, Inc. (2017)
31. Pandey, S., Tokekar, V.: Prominence of MapReduce in big data processing. In: 2014 Fourth International Conference on Communication Systems and Network Technologies, pp. 555–560 (2014)
32. Piras, M.E., Pireddu, L., Moro, M., Zanetti, G.: Container orchestration on HPC clusters. In: Weiland, M., Juckeland, G., Alam, S., Jagode, H. (eds.) High Performance Computing, pp. 25–35. Springer International Publishing, Cham (2019)
33. Plauth, M., Feinbube, L., Polze, A.: A performance survey of lightweight virtualization techniques. In: De Paoli, F., Schulte, S., Broch Johnsen, E. (eds.) Service-Oriented and Cloud Computing, pp. 34–48. Springer International Publishing, Cham (2017)
34. Prabhakaran, S.: Dynamic resource management and job scheduling for high performance computing. Ph.D. thesis, Technische Universität Darmstadt, Darmstadt, Aug 2016
35. Priedhorsky, R., Randles, T.: Charliecloud: unprivileged containers for user-defined software stacks in HPC. In: Proceedings of the International Conference for High Performance Computing, Networking, Storage and Analysis, New York, NY, USA, SC 17. Association for Computing Machinery (2017)
36. Reuther, A., Byun, C., Arcand, W., Bestor, D., Bergeron, B., Hubbell, M., Jones, M., Michaleas, P., Prout, A., Rosa, A., Kepner, J.: Scalable system scheduling for HPC and big data. J. Parallel Distrib. Comput. **111**, 76–92 (2018)
37. Rodriguez, M.A., Buyya, R.: Container-based cluster orchestration systems: a taxonomy and future directions. Softw. Pract. Exper. **49**(5), 698–719 (2019)
38. Rosado, T., Bernardino, J. An overview of openstack architecture. In: Proceedings of the 18th International Database Engineering & Applications Symposium, New York, NY, USA, IDEAS '14, pp. 366–367. Association for Computing Machinery (2014)
39. Senthil Kumaran, S.: Practical LXC and LXD: Linux Containers for Virtualization and Orchestration, 1st edn. Apress, USA (2017)
40. Soppelsa, F., Kaewkasi, C.: Native Docker Clustering with Swarm. Packt Publishing (2017)
41. Staples, G.: Torque resource manager. In: Proceedings of the 2006 ACM/IEEE Conference on Supercomputing, New York, NY, USA, p. 8. Association for Computing Machinery (2006)
42. Vavilapalli, V.K., Murthy, A.C., Douglas, C., Agarwal, S., Konar, M., Evans, R., Graves, T., Lowe, J., Shah, H., Seth, S., Saha, B., Curino, C., O'Malley, O., Radia, S., Reed, B., Baldeschwieler, E.: Apache hadoop YARN: yet another resource negotiator. In: Proceedings of the 4th Annual Symposium on Cloud Computing, New York, NY, USA, SOCC '13. Association for Computing Machinery (2013)
43. White, T.: Hadoop: The Definitive Guide. O'Reilly Media, Inc. (2012)
44. Wrede, F., and von Hof, V. Enabling efficient use of algorithmic skeletons in cloud environments: container-based virtualization for hybrid CPU-GPU execution of data-parallel skeletons. In: Proceedings of the Symposium on Applied Computing, New York, NY, USA, SAC 17, pp. 1593–1596. Association for Computing Machinery (2017)

45. Xavier, M.G., Neves, M.V., Rossi, F.D., Ferreto, T.C., Lange, T., De Rose, C.A.F.: Performance evaluation of container-based virtualization for high performance computing environments. In: 2013 21st Euromicro International Conference on Parallel, Distributed, and Network-Based Processing, pp. 233–240 (2013)
46. Younge, A.J., Pedretti, K., Grant, R.E., Brightwell, R.: A tale of two systems: using containers to deploy HPC applications on supercomputers and clouds. In: 2017 IEEE International Conference on Cloud Computing Technology and Science (CloudCom), pp. 74–81 (2017)
47. Zaharia, M., Chowdhury, M., Franklin, M.J., Shenker, S., Stoica, I.: Spark: cluster computing with working sets. In: Proceedings of the 2nd USENIX Conference on Hot Topics in Cloud Computing, USA, HotCloud10, p. 10. USENIX Association (2010)
48. Zaharia, M., Xin, R.S., Wendell, P., Das, T., Armbrust, M., Dave, A., Meng, X., Rosen, J., Venkataraman, S., Franklin, M.J., Ghodsi, A., Gonzalez, J., Shenker, S., Stoica, I.: Apache spark: a unified engine for big data processing. Commun. ACM **59**(11), 56–65 (2016)
49. Zhang, J., Lu, X., Panda, D.K.: Is singularity-based container technology ready for running MPI applications on HPC clouds? In: Proceedings of The10th International Conference on Utility and Cloud Computing, New York, NY, USA, UCC 17. Association for Computing Machinery (2017)
50. Zhou, N., Georgiou, Y., Zhong, L., Zhou, H., Pospieszny, M.: Container orchestration on HPC systems. In: 2020 IEEE International Conference on Cloud Computing (CLOUD) (2020)

# Trends in HPC and AI

# The Role of Machine Learning and Artificial Intelligence in High Performance Computing

**Michael M. Resch and Bastian Koller**

**Abstract** High Performance Computing has recently been challenged by the advent of Data Analytics (DA), Machine Learning (ML) and Artificial Intelligence (AI). In this paper we will first look at the situation of HPC which is mainly shaped by the end of Moore's law and an increase in electrical power consumption. We then explore the role that these technologies can play when coming together. We will look into the situation of HPC and into how DA, ML and AI can change the scientific and industrial usage of simulation on high performance computers. Finally, we make suggestions of how to use the convergence of technologies to solve new problems.

## 1 Introduction

Machine Learning (ML) and Artificial Intelligence (AI) have become more visible over the last years and have developed into fields that show a huge potential for using computers in a variety of applications. Areas of usage range from improving and speeding up medical image processing, to optimizing urban planning processes, to a standardized and high speed handling of banking processes even in the usually heavily personalized consumer market. Some economists assume that ML and AI will change the world so dramatically that millions of jobs will be lost and we need to speak of a "second machine age" [1]. But this is not the scope of a scientific investigation.

In this article we have a look at the merger of High Performance Computing (HPC) with ML and AI. The situation of HPC has been described before [2, 3] and is considered to be interesting but also limited by the technical problems that we face with the end of Moore's law. We will argue that ML and AI have to be seen as two different technologies that are part of a chain of technologies that naturally lead us from HPC to AI. We will also argue that HPC is not only a facilitating technology for ML and AI but also a heavy user of ML and AI technologies. In this paper we will

M. M. Resch (✉) · B. Koller
High Performance Computing Center Stuttgart (HLRS), University of Stuttgart, Nobelstrasse 19, 70569 Stuttgart, Germany
e-mail: resch@hlrs.de

 
M. M. Resch et al. (eds.), *Sustained Simulation Performance 2019 and 2020*,
https://doi.org/10.1007/978-3-030-68049-7_11

not address ethical issues that come with ML and especially AI. This is an important topic but should not be part of a technical overview paper.

Finally, we give an outlook of how HPC, ML and AI will be merged into a new world of technologies. This merger will allow to tackle new research questions and will help to substantially improve our way of doing simulation on supercomputers.

## 2 The Status of High Performance Computing

The future of HPC is widely discussed in the scientific community but has also attracted substantial political interest in the last years. The topic around which these discussions evolve is the race for the first Exaflop (10^18 floating point operations per second) system. The US has announced to build such systems in the coming years [4]. Japan has announced to build a system as a follow-on national project for the current RIKEN K-Computer system [5]. China has started a program to build such a system and is planning to have up to eight systems, each capable of Exaflop performance [6]. Europe has started an initiative called EuroHPC [7] which has decided to fund three European pre-Exaflop systems in 2021 and plans to fund two Exaflop systems later on. All in all, the world of HPC seems to be set for moving from the era of Petaflops (10^15 floating point operations per second) computing to the new era of Exaflops computing smoothly and without much trouble.

Having a look at the list of the fastest supercomputers in the world [8], we see that in June 2020 four systems can be considered to be already in the pre-Exaflop range. The fastest system in the world (Fugaku) shows a peak performance of about 500 PFLOPS and a Linpack performance of more than 400 PFLOPS. Assuming a growth rate as expected by Moore's law [9] we should see a performance increase of two every 18 months. That would be a 1 EFLOPS system in November 2021. Assuming that Moore's law still holds and assuming that the budget of 500 million US$ is an increase in system cost of about 50% compared to the Summit system, an Exaflop system seems to be at least feasible in 2021.

It becomes obvious in this calculation that the main commonality of all international projects aiming at Exaflop systems is a substantial increase in budget. One reason is that the total energy costs for a top 10 system have increased over the last decade and are still increasing. For an Exaflop system to be operational in the year 2021 we expect to see a power consumption in the range of 30–50 MW. Costs for electricity vary from country to country but such a high power consumption substantially increases total cost of ownership. On the other hand, the increase in investment is necessary to make up for the slow-down in Moore's law as computers gain speed mainly by increasing the number of processors used.

# 3 Slowing Down the Speedup

Even though the question of architecture is perhaps no longer the most pressing one, the first problem that HPC will have to handle in the future is a hardware problem: the end of Moore's law [9]. The prediction of Moore in 1965, that we would be able to cram ever more transistors on the same surface, with the numbers doubling every 12—later Moore shifted this to 18—months, did hold for about 50 years. As of today, it is unclear how far we still can get with miniaturization. Seven nanometres are already used in manufacturing. Five nanometres and finally three nanometres might be achievable in the foreseeable future. However, it is unclear whether this is also economically feasible. Already today there are only three manufacturers that handle the 7 nm process (Samsung, TSMC, Intel). As a consequence, we need to assume that beyond the mid-2020s increase in performance for supercomputers will be difficult to achieve through a further increase in transistors on a chip [10].

## *3.1 Quantum Computing*

This triggers the need for new solutions. Quantum Computing has started to carry the hopes of funding agencies to the point that departments and divisions are renamed to be responsible for "Quantum Computing and HPC". Given that there are no real quantum computers available yet, this paper will not cover the topic of quantum computing in detail. What we see when we look at first systems that are similar to quantum computers is that these systems will most likely not provide higher performance in the sense of more floating point operations but will rather open up a new field of simulation.

Two key question seems to be important from the point of view of HPC for Quantum Computing as of today.

1. Is there a simple way to transform classical HPC simulation problems into equivalent problems to be solved by quantum computers? A first investigation indicates that it might be possible to find a way to map minimization problems to the problem of solving a system of equations. But this will need further investigation.
2. Will users accept the fact that with quantum computing the notion of a deterministic solution is lost or at least weakened? This might become a question of interpretation of results and may not be such a big issue, given that already today for many simulations uncertainty quantification has to be provided.

Quantum Computing will definitely remain a field of research for the coming decades. It will take some time before productive systems can be made available for scientific usage and still more time to turn them into a widely spread device to be used in science and industry alike.

### *3.2 Software, Methods and Algorithms*

With traditional hardware making it much more difficult to squeeze more performance from a given HPC architecture, the focus of attention will have to be shifted towards software, mathematical methods, and algorithms for HPC simulation. Various investigations show that over the last decades, mathematical methods—at least for the solution of systems of equations—have already contributed substantially to the overall increase in simulation performance [11].

This will not only have to be continued in the future. Much more than that, software, algorithms, and mathematical methods will take over as the main driving force in getting more performance and better results from future machines. The slowing down of Moore's law will take away some of the steam in HPC that came from the impressive hardware performance increase. On the other hand, with slowing performance increase it is most likely that HPC systems will extend their individual life time. While today an HPC system lasts for about 3–5 years it may well last about 5–8 years in the future. This will allow software developers to spend more time optimizing for existing hardware. In turn scientific software will be able to better exploit performance features of modern architectures. But also commercial software will benefit as independent software vendors have a higher chance to get a reasonable return of investment when optimizing software for a given hardware platform.

The focus of attention for future optimizations in software, algorithms, and mathematical methods is twofold. First, an increase in compute performance has to be achieved as modern HPC systems exhibit a lamentably low level of sustained performance for average applications. Second, all new methods will have to keep in mind power consumptions.

First investigations show that optimized codes can run at a substantial lower level of power consumption when power consuming behaviour can be avoided. The good news is that we find that for both targets optimum usage of memory is the key. The access to slow main memory both slows down computation and increases power consumption. Approaches that are aware of the memory wall can hence help to solve two problems of high performance computing at the same time. Finding such approaches will require a deep understanding of hardware and software and will need a lot of data.

## 4 From Big Data to Artificial Intelligence

Another trend that has a huge impact on HPC is what can best be described as the transition from "Big Data" through Machine Learning towards Artificial Intelligence. Even though the current usage goes way beyond the original idea of mainly handling large amount of data the term "Big Data" still in a sense is useful as it describes well how HPC may be overwhelmed by the data problem in the coming years. HPC may well become a smaller part of a larger digital infrastructure that is focusing around data rather than around compute power. We will address, how this will impact HPC.

Over the last years a number of new paradigms and one old paradigm have grown in importance. All of these are based on data. Big Data was already introduced more than a decade ago and for a while was considered to bring a new paradigm to science [12]. Some even went further to claim that with Big Data science would reach the "end of theory" [13]. However, correlation and causality are two different things and hence the simple analysis of data without a theory will always show correlation but never causality.

Big Data was soon further developed into Data Analytics and then further into a concept that brings together data and insight, and which is usually called Machine Learning. But at the same time a new wave of Artificial Intelligence projects has hit the high performance computing community. From an HPC expert point of view there is a logical path from Big Data to Artificial Intelligence that can be seen as a new chance for simulation on HPC. In the following we will describe how we can find a continuous spectrum of applications ranging from classical simulation to Artificial Intelligence.

### *4.1 Classical Simulation*

In the classical simulation approach the simulation expert goes through a series of steps which are usually handled sequentially. We ignore here input data and focus on output data. The results of a simulation are usually analysed post mortem in a visualization environment. So, data are created and each data set is considered individually. Visualization provides the necessary techniques for analysis. Usually all simulation runs are independent. The simulation expert has a clear understanding of the job she is running and also knows which features or values to look for in the computed results. For a Computational Fluid Dynamics (CFD) simulation this usually means to look for velocities and pressure and to visually identify spots where special flow phenomena—like turbulence, recirculation, stagnation—appear [14]. A global view of all simulation runs or a deep dive into the data is usually not undertaken. However, what seems to be simple is a highly complex problem as visualizations have become four-dimensional over time and hardware environments for visualizations have become a technology of its own. Nevertheless, post mortem visualization is a relatively simple process as it mainly intends to look at data.

### *4.2 Big Data in HPC*

The concept of Big Data evolves from traditional analysis of data and looks at data from the point of view of harvesting information that may be buried and hidden in too many data for human beings to analyse. HPC simulation is currently moving from traditional simulation to Big Data in the sense that simulations create huge amounts of data. These data can still be visualized but the human eye is unable to grasp all details and to identify all interesting spots. The promise of the "end of theory" [13] will most probably not materialize in HPC simulation as analysing simulated data requires a deep understanding of the overall simulation process. However, concepts of Big Data may help to create awareness in the HPC community that classical visualization methods may not be enough to fully exploit the knowledge created by an HPC simulation. For our CFD example Big Data may lead the simulation expert to explore several simulation results at a time. It may also make the user want to start to automatically search for features (stagnation, turbulence, vortices, …) based on improved evaluation methods.

## 4.3 *Machine Learning in HPC*

Machine Learning is a technology that not necessarily evolves from Big Data. However, it can be seen as a logical continuation of the idea to extract information from large amounts of data. If we assume that we still need some theory to extract knowledge from data, we need to be able to use the data we have to improve the theory. The learning process, however, now goes beyond the pure analysis of data. It makes use of the data to improve our understanding and lead us to improved or new theories. When we now look at our example from flow simulation, machine learning can help to use existing simulation data to learn how to design future simulation runs or to learn how to interpret a large number of simulation results in a coherent view.

## 4.4 *Artificial Intelligence in HPC*

The notion of Artificial Intelligence (AI) is said to have been first introduced by Alan Turing back in 1950 [15]. Intelligence is a concept that is basically not a technical one. Over the last decades it has seen a change in meaning and understanding. It is hence a bit difficult to clearly judge the technical merits of AI. While Alan Turing was referring to AI as a computer system that is able to fully imitate the logical behaviour of a human being, modern interpretation of AI is looking at two main features.

On the one hand, AI is considered to be a way to create humanoid robots (imitating human behaviour). The focus of this approach is to create an artificial human being including the physical body. This has little to do with HPC per se and will not be further considered here.

On the other hand, AI is considered to be able to replace human beings in the decision making process. During the 1970s and the 1980s there was substantial investment in AI research and expectations to achieve both goals were high. The most recent wave of enthusiasm about AI has a more realistic focus. It usually aims at integrating software and hardware solutions with enough data to create a system that is able to unburden the human being from complex but standardized decisions. Typical examples are decision making in medical treatment and decision making in the analysis of human faces. This is certainly far away from the original human-like machine. However, the potential for this technology is high. When we come back to our CFD example AI can help to learn from previous simulations to make decisions about the future simulations that have to be done to solve a given problem. The decision making in the simulation would practically entirely be offloaded to an AI system.

What we see when looking at these three technologies are two things:

- There is no clear distinction to be made between HPC, Big Data, ML and AI. These technologies are a gradual advancement from a process purely controlled by the human being towards a process nearly entirely controlled by what we might call machine intelligence.
- HPC is not a technology separate from Big Data, ML and AI but all these technologies rely heavily on the availability of both compute power but also theoretical knowledge.

## 5 What Does This Mean for HPC?

Even though a traditional look at HPC already shows some dramatic change, there is something that might be even more important for HPC. Considering current trends, we find that HPC is going to be part of something bigger—which is driven by data but not only data. It is meanwhile well accepted that there is value in the data. However, there is much more value in the right learning processes and algorithms.

HPC simulations might be one source for such data. Sources of data can however be manifold:

- The traces each person is generating each day using systems in the internet, when shopping, when communicating, when watching movies, when visiting other web pages.
- The data of business operations which are digitally available and stored for years.
- The increasing amount of sensors everywhere especially powered by the Internet of Things (IoT), going from production lines to personal homes—smart meters are a good example for that.

Two main scenarios for HPC in such a data driven world evolve and are briefly discussed here.

## 5.1 HPC Needs Data

HPC simulation will increasingly make use of modern methods to handle, explore, interpret, and turn data into decisions. The simulation community will move from classical batch processing or co-simulation with visualization and simulation running in parallel towards a setup that is driven by data. Simulations will bring in more data from fields other than simulation. Meteorology is an example where measured data combine with simulation data in order to improve the quality of the picture.

Another example is pandemic simulations which have to take into account data collected from hospitals and doctors and have to update models every day with these data. In that case simulation uses data to aim at re-simulating the development of the pandemic over the last days or hours. At the same time these backward-simulations improve parameter estimates to do forward simulations in order to understand the evolving pandemic.

Simulations will also bring in data analytics methods in order to better understand computed results. This will take away control of visualization from the human being and put it more in the "hands" of the computer system. But the change will go even further. AI systems will help to analyse simulation runs and learn from the results in order to make suggestions for future simulations. In a mid-term perspective, simulations could even be entirely taken off the hands of human beings and be done by AI systems that access simulation data and theory repositories automatically responding to user questions through simulation and their interpretation. As strange as this may sound to traditional simulation experts it would only be a continuation of a process in which the behaviour of computers is hidden from the user. And it would be the logical evolvement of all technology that is supposed to replace human beings in order to improve and/or speedup a process that can be standardized.

## 5.2 Data Needs HPC

When looking carefully at the requirements and the potential of data analytics, ML and AI it is obvious that these technologies will not replace HPC but will rather give a new boost to HPC. One of the key aspects in ML and AI is the learning phase. While it is obvious that data are required to learn it is less obvious that compute power is a must for this learning phase. It is hence not surprising that the fastest Japanese supercomputing system in November 2019 [16] was exclusively devoted to Artificial Intelligence. The AI Bridging Cloud Infrastructure (ABCI) has a focus on applications from AI and will serve the Japanese research and industrial community for the coming years. In that sense HPC will have a new user community that will increase the need for large scale systems and Exaflop performance. The compute needs of this community are currently exploding. With the usage of general purpose graphics processing units (GPGPUs) in supercomputers, new high performance computers are well prepared for the challenge.

## 6 Who Might Benefit?

We can find a number of interesting cases that will benefit from a merger of HPC with data technologies. Some of them are rather obvious. Others do not seem to be good candidates in the first place.

Banks are one potential group of customers that may move even further into the field of HPC. They already have a history of analysing data when it comes to stock exchange analysis. There are further topics that might be interesting. Fraud detection is one field that might benefit both by increasing the speed of a detection but also by increasing the level of accuracy. Permanent and individualized portfolio analysis both for institutional and private customers is a field that will need HPC performance.

In many of the business cases where the analytics is done in large in-memory data bases, not many are thinking about HPC. However, after the analysis of business data a next step would be to change and improve the situation. In several cases this could require of large simulations and parameter studies which will naturally require HPC systems. A good example for this is railroad companies. In case of delays simulations are used to decide between different options to improve the difficult traffic situation.

The increasing use and number of linked sensors is another area where data volumes are exploding. This leads to the idea of in-time analytics to detect events before they actually occur for example with machine learning technologies. This may lead to new insights and better understanding of existing dependencies. In order to extract such information inverse problems will have to be solved. This will require HPC to a much bigger extent than today.

Another example with even higher impact on HPC is the usage of sensors to detect major natural disasters which might lead to damage and loss of lives. In case of a marine earth quake Japan has set up a system to automatically analyse data, simulate the impact of a tsunami and take measures to protect its people. This is an example where data, simulation and AI have to work together to come up with a solution that could never be achieved with classical simulation approaches. Given the time-critical situation and at the same time the financial impact and the risk for human lives only such an integrated approach can help to come to acceptable solutions.

## 7 What Does This Mean for HPC Environments and Architectures?

The development described above already has an impact on architectures and overall HPC environments. I/O and the handling of large data sets are considered to be a critical topic in HPC procurements and in systems offered. Specialized I/O nodes are part of any HPC system already today. They will become more important in the future. Large memory nodes to be able to handle larger data sets have become a standard component. The size of memory is continuously increasing. More and more GPGPU-nodes are integrated in HPC-systems.

In several cases a direct connection to include up-to-date input data into the ongoing simulations will require a change in the HPC environment setup and will require to solve new security issues. Additionally, there is the upcoming requirement for "urgent" computing which needs to be solved administratively as well as technically as many HPC systems are not prepared for such a requirement. The main problem for HPC operations in urgent computing is the fact that running jobs will have to be interrupted such that some users may lose their jobs or results.

## 8 Conclusion

Summarizing our findings, we see a number of trends which will have an impact on HPC and AI in the coming years. It is getting ever more clear that the main driving force of HPC in the last decades will go away. Moore's law is coming to an end and will not help us increase HPC performance in the future. Improved algorithms and mathematical methods will still have the potential to increase sustained performance but will only extend the race in HPC without being able to overcome the stagnation in peak performance to be expected. Power consumption is a new problem which will be with the HPC community for at least the next decade.

At the same time, we see a shift away from pure HPC to an integration of technologies. Big Data, Machine Learning and Artificial Intelligence are added to the set of tools that help to solve many of the traditional problems much better and to be able to tackle new problems. This convergence of technology will for the coming ten years be the most important aspect in High Performance Computing.

## References

1. Brynjolfsson, E., McAfee, A.: The Second Machine Age: Work, Progress, Prosperity in a Time of Brilliant Technologies. W. W. Norton & Company (2016)
2. Resch, M.M., Boenisch, T., Gienger, M., Koller, B.: High performance computing—challenges and risks for the future. In: Singh, V.K., Gao, D., Fischer, A. (eds.) Advances in Mathematical Methods and High Performance Computing. Springer (2019)
3. Resch, M.M., Boenisch, T.: High performance computing—trends, opportunities and challenges. In: Ivanyi, P., Topping, B.H.V., Varady, G. (eds.) Advances in Parallel, Distributed, Grid and Cloud Computing for Engineering, pp. 1–8. Saxe-Coburg Publications (2017)
4. https://www.energy.gov/articles/us-department-energy-and-intel-build-first-exascale-supercomputer. Accessed 20 Nov 2019
5. https://www.r-ccs.riken.jp/en/postk/project. Accessed 20 Nov 2019
6. Private communication with Chinese colleagues (2019)
7. https://eurohpc-ju.europa.eu/. Accessed 20 Nov 2019
8. www.top500.org. Accessed 20 Nov 2019
9. Moore, G.E.: Cramming more components onto integrated circuits. Electronics **38**(8), 114–117 (1965)
10. Courtland, R.: Transistors could stop shrinking in 2021. IEEE Spectrum. https://spectrum.ieee.org/semiconductors/devices/transistors-could-stop-shrinking-in-2021. Accessed 20 Nov 2019

11. Marra, V.: On solvers: multigrid methods. https://www.comsol.com/blogs/on-solvers-multigrid-methods/. Accessed 20 Nov 2019
12. Hey, T., Tolle, K.M., Tansley, S.: The Fourth Paradigm: Data-Intensive Scientific Discovery. Microsoft Research (2009)
13. Anderson, C.: The end of theory: the data deluge makes the scientific method obsolete. Wired Mag. (2008)
14. Perktold, K., Resch, M., Peter, R.: Three-dimensional numerical analysis of pulsatile flow and wall shear stress in the carotid artery bifurcation. J. Biomech. **24**(6), 409–420 (1991)
15. Turing, A.M.: Computing machinery and intelligence. Mind **59**, 433–460 (1950)
16. https://www.top500.org/system/179393. Accessed 20 Nov 2019

# Trends and Emerging Technologies in AI

**Dennis Hoppe**

**Abstract** The growth of artificial intelligence (AI) is accelerating. AI has left research and innovation labs, and nowadays plays a significant role in our everyday lives. The impact on society is graspable: autonomous driving cars produced by Tesla, voice assistants such as Siri, and AI systems that beat renowned champions in board games like Go. All these advancements are facilitated by powerful computing infrastructures based on HPC and advanced AI-specific hardware, as well as highly-optimized AI codes. In this paper, we will thus overview current and future trends in AI, as well as emerging technologies that drive AI innovation. We will spend a significant part on the ethical aspects that arose around AI whenever citizens interact with AI systems. This translates directly to key topics such as transparency, trustworthiness, and explainability of AI systems. This paper will therefore discuss several approaches of the new research field called explainable AI (XAI). Finally, we will present briefly AI-specific hardware that may find its way into HPC computing.

## 1 Introduction

Both domains, artificial intelligence (AI) and high-performance computing (HPC), will ultimately have to fully converge to support AI workflows on HPC. Although it should be evident that AI needs to harness the pure processing power of today's and future HPC infrastructures, HPC needs first to be made "ready" for AI. However, this is another topic, and thus the diverse and very dynamic requirements of AI, which typically clash with the legacy software stack of HPC, will be discussed at another time; we are referring the interested reader to these publications [1–5]. Instead of discussing how to break down technical barriers, this paper focuses on trends and emerging technologies in AI that have relevance for HPC, too.

In the last years, new research fields emerged or were revived around several themes: ethical aspects of artificial intelligence, automation in industrial facilities, automation of machine learning and deep learning workflows, personal medicine,

D. Hoppe (✉)
Höchstleistungsrechenzentrum Stuttgart, Nobelstraße 19, 70569 Stuttgart, Germany
e-mail: hoppe@hlrs.de



M. M. Resch et al. (eds.), *Sustained Simulation Performance 2019 and 2020*,
https://doi.org/10.1007/978-3-030-68049-7_12

speech recognition, meta-learning and reinforcement learning, to name but a few. Aren't we forgetting something? Quantum computing (QC); specifically hybrid solutions [6, 7] leveraging today's power of HPC in combination with QC yielding solutions that are immediately applicable, give quantum computing a significant boost. These themes can be found also in the top 10 AI trends identified for 2020 [8].

## *1.1 The Hype Cycle Model*

If we take a closer look at the annual Gartner Hype Cycle for AI from 2019, as depicted in Fig. 1, we see several reoccurring themes [9]: reinforcement learning [10, 11], explainable AI [12–14], edge AI [15, 16], AutoML [17–19]. The hype cycle follows a model introduced by [20], where expectations (y-axis) are associated with a given innovation over time (x-axis). The hype of an innovation then follows five stages [20, 21]:

1. **Innovation trigger** signals the breakthrough of a new innovation. Innovation comes often from academia or research and development departments (e.g.., startups). This phase is driven by early adopters. Current innovations in this phase: reinforcement learning; explainable AI.
2. **Peak of inflated expectations** reassures the potential value of an innovation and thus leads to a wider adoption of a new technology by additional stakeholders. Current innovations in this phase: AutoML.
3. **Trough of disillusionment** describes a period of realization that the innovation can not fulfil all expectations, and thus the the applicability of the innovation needs to be reviewed. Number of adopters is at its lowest. Current innovations in this phase: FPGA accelerators and autonomous driving.
4. **Slope of enlightenment** focuses on overcoming barriers observed in the trough and finding new value. This phase is driven by a second generation of products based on an innovation in question; this phase also benefits from best practices and lessons learned so far. Based on the last hype cycle from 2019, there are currently no innovations in this phase. We will, however, see autonomous driving and virtual assistants like Alexa or Cortana quite soon in this phase [22].
5. **Plateau of productivity** finally describes mature solutions of a given innovation applied in real-world leading to a significant higher adoption of the new technology. Current innovations in this phase: GPU accelerators and speech recognition. Based on the fact that speech recognition has achieved high maturity, virtual assistants will closely follow.

Besides trends in AI in general, we also see a vast amount of emerging technologies around AI-specific hardware. Whereas recent years were dominated by accelerators such as GPUs, FPGAs, and ASICS, we now also see new developments such as TPUs and all technologies that are pushed by edge AI. Edge computing comes with dedicated requirements such as low-power and a strong focus on inference instead of support for training AI models [23].

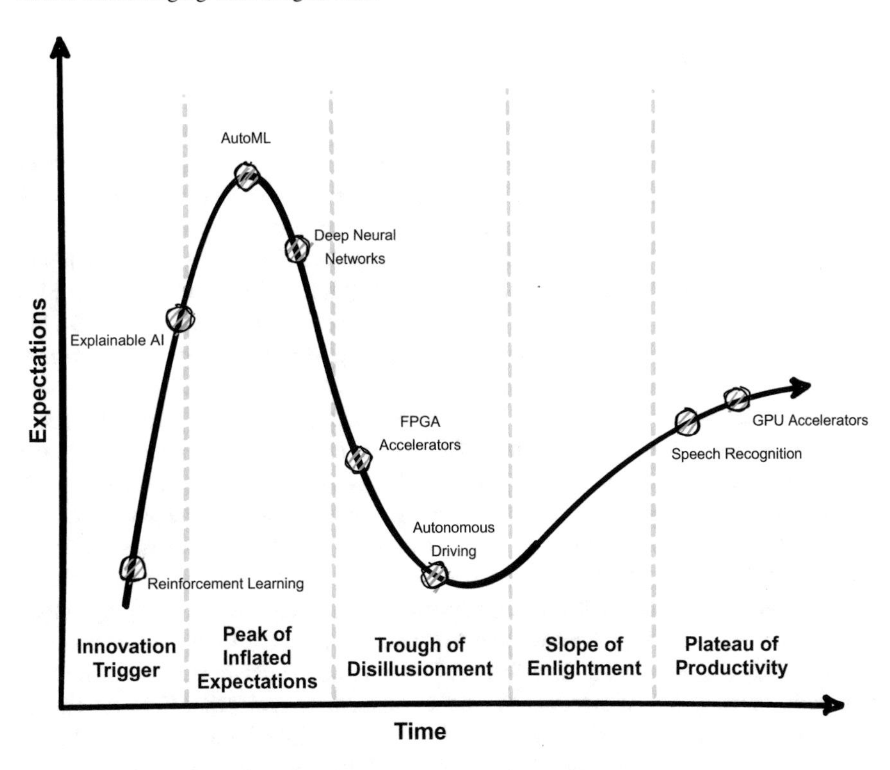


**Fig. 1** Gartner hype cycle for AI technologies as of 2019 [9]

## *1.2 Structure of This Paper*

The remainder of this paper is organized as follows: Sect. 2 introduces the needs for addressing ethical aspects of AI, the societal impact of AI, as well as lists current approaches towards explainable AI such as decision trees or dedicated methods such as LIME and SHAP. Section 3 then continues in discussing trends in AI related to software and algorithms, specifically mentioning the domain of automated machine learning and meta-learning, which has seen a striking raise of interest in recent years. Section 4 then complements the survey by highlighting emerging technologies in AI with a focus on hardware developments, mentioning GPUs as well as requirements imposed by edge computing ("edge AI"). This paper finally concludes with a summary and future outlook.

## 2 Ethical Aspects of AI

The main aim of this chapter is to answer what explainable AI is, what it is trying to solve, and what existing and new techniques exist to achieve explainable AI workflows. But before answering these questions, let's take a step back and look at the big picture. With the advent of AI, specifically in the last decade, we became aware of the trans-formative impact AI has onto society. Here, we are deliberately leaving out the discussion on the effect that AI will have on the automation of jobs and thus changing and shaping the future workforce. We will rather focus on the rising challenge of trusting decisions made by AI algorithms (e.g., neural networks) and systems as a whole (e.g., interacting with virtual assistants). Studies have shown, that society tends to put too much trust into decisions made by AI systems [24, 25], e.g., into the autopilot capabilities of self-driving cars. Thus, in order to increase the adoption of AI, algorithms and systems should follow some basic guidelines such as introducing transparency to decision making. For example, a self-driving car may inform about hazardous situations and warns drivers when new scenarios are faced that can not be handled by the AI system itself and need human interaction. Another example would be that a self-driving car informs a driver that it will change lanes in a few seconds because it would like to overtake the car in front. In this context, another study suggests that more effort needs to put into mental modeling research, so that both, humans and AI systems, can better understand each other [24]. An expert group on AI recently coined the term Trustworthy AI, which is based on three building blocks: Lawful AI, Ethical AI, and Robust AI. Enclosed are seven requirements that a trustworthy AI system should meet:

- human agency and oversight,
- technical robustness and safety,
- privacy and data governance,
- diversity and fairness,
- societal and environmental well-being, and
- accountability.

We may refer the interested reader to the ethic guidelines of the European Commission for more insights [14]. Here, we would like to come back to the transparency aspects of ethical AI, which brings us to the field of explainable AI. Another framework is proposed by the AI Ethics Impact Group led by VDE [26]. The authors follow a traffic-light based rating system, as shown in Fig. 2, to give an AI system an ethics label. The rating system assigns each dimension, e.g., transparency, a rating from A (high) to G (low). Furthermore, each rating is based on a set of criteria (e.g., disclosure of data sets), indicators (e.g.., documentation of data) and so-called observables (e.g., "yes, data is documented"), making the rating system itself transparent.

**Fig. 2** Labelling suggestion for explainable algorithms and systems based on 7 levels ranging from A to G (as seen in [26]). Here, level C is selected with respect to the transparency aspects (as seen in [26])

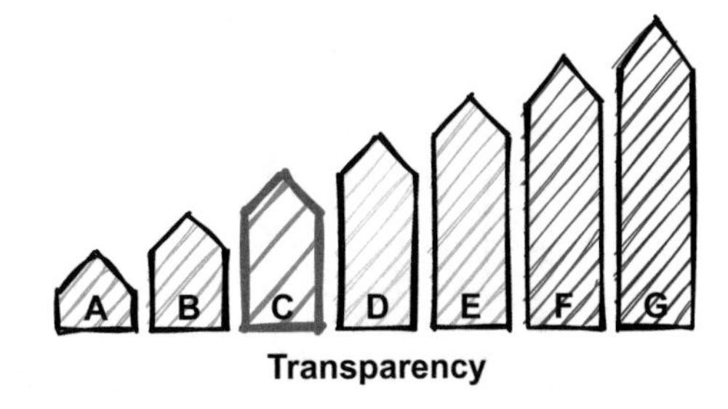


## 2.1 *Explainable AI (XAI)*

Besides developing these specific frameworks for grading AI systems with respect to their ethical aspects, there is also research conducted on the algorithmic-level itself to make algorithms more transparent and explainable. The rather new field of explainable AI traces back to the fact that most AI algorithms are fundamentally black boxes [12, 27]. The issues lies deep in the fact that vast amounts of data are processed and then mapped onto predicting classes and outcomes (e.g., a score for the current health status), but the underlying reasons why this outcomes was obtained, are not exposed. Although neural networks have demonstrated their effectiveness in image recognition and processing in general, the way they are doing it and how they produce and present results to humans is often vague and can not be explained [28]. Therefore, this section will give first an overview about classical techniques from machine learning (interpretable models) to improve the explainability of algorithms, and then we discuss some new approaches such as LIME and SHAP, which act as surrogates to replace more complex models with simpler ones. The structure of this section loosely follows the overview given in [29].

### 2.1.1 Classical Models

The models are based on well-known algorithms used in everyday machine learning: linear regression, decision trees, and neural networks. We will first briefly describe each algorithm, and then discuss some advantages and disadvantages. We assume that the reader is familiar with the basic concepts of machine learning.

#### Linear Regression

***Basics*** Linear regression is a statistical technique to describe the relationship between one (dependent) variable with two or more (independent) variables. We try for a given variable $x$, a so-called feature, to predict a real-world value $y(x)$ using the following linear model function:

$$y(x) = \omega_0 + \sum_{i=1}^{p} \omega_j \cdot x_j,$$

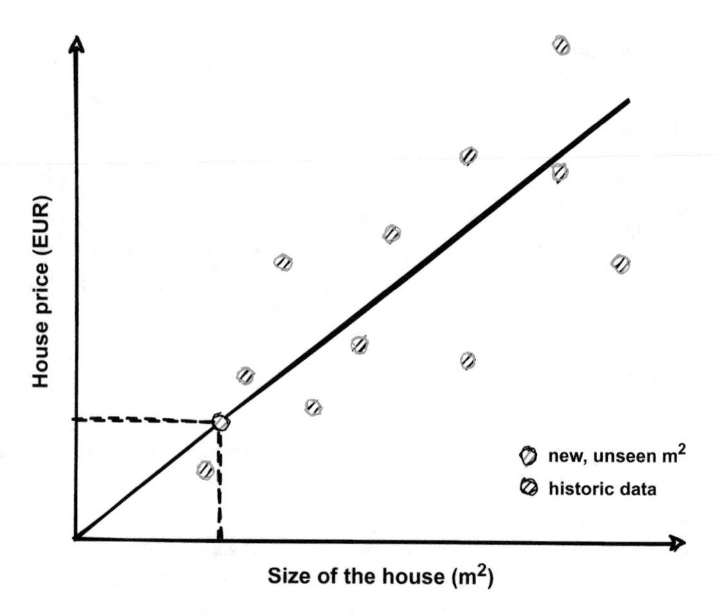


**Fig. 3** Prediction of house prices based on the independent variable `square meter size` using linear regression. Based on historic data that includes a mapping between square meters and house prices, a straight line is fitted to the historic data that minimizes the error. A new data point (here: $m^2$ size) can then be mapped to a house price (cf. dashed lines)

where $\omega$ are weights (i.e., the slope of the straight line) and $y$ represents the ground truth. A common example that is used to explain linear regression is the prediction of house prices based on the dependent variable "size of the house in square meters" and the independent variable "house price", the one that needs to be predicted (cf. Fig. 3).

It is then possible to evaluate the so-called goodness of fitting the data by computing the residual sum of squares (RSS) as follows:

$$RSS(\boldsymbol{\omega}) = \sum_{i=1}^{n}(y_i - \boldsymbol{w}^T \boldsymbol{x}_i)^2,$$

where $\boldsymbol{x} = (x_1, \ldots, x_n)$ is the vector of the feature space $X$, and $\boldsymbol{\omega} = (\omega_1, \ldots, \omega_p)$ consequently. Thus, there is a direct evaluation measure to assess how good the linear model fits the original data. Further, we have the possibility to rank each feature based on its importance, meaning how much influence it has towards the prediction: e.g., the $t$ test [30].

***Discussion*** The linear regression is applied often in economics and health because the mathematics behind it is straight forwarded, and a means for feature importance and the actual quality of fitting the data is given. Moreover, a lot of visualization exist that help to interpret results. In the example of house price prediction, it becomes

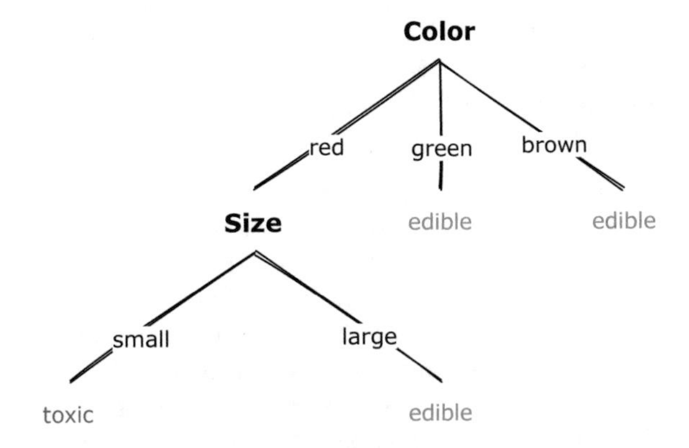


**Fig. 4** Visualization of the decision tree for the mushroom example (cf. Table 1) with two attributes: color and size. This tree successfully classifies examples given into toxic and edible mushrooms

**Table 1** Table with examples for edible and non-edible mushrooms. Decision is made based on two attributes: color and size

| | Color | Size | Edible |
|---|---|---|---|
| 1 | red | small | toxic |
| 2 | brown | small | edible |
| 3 | brown | large | edible |
| 4 | green | small | edible |
| 5 | red | large | edible |

somewhat obvious that higher house prices are linked to larger living space. As a downside, the applicability of linear regression is limited, because a linear model often oversimplifies the relationships between variables.

### Decision Trees

***Basics*** Going away from predicting a concrete value,[1] let us know have a look at a classification technique named decision trees. The basic idea behind decision trees is that they follow simple if-else rules in order to traverse a (binary) tree built from variables (cf. our feature space) that ultimately map to a specific class, which are located at the leaves of a tree. Each internal node then basically represents a decision to be made that eventually enables us to make a classification decision after traversing the entire tree. It is obvious that the higher the height of tree, the more complex the decision rule gets.

Figure 4 depicts a simple decision tree that classifies mushrooms into toxic and eatable ones (as seen in [31]); the available feature space is composed of the attributes *color* and *size*. Table 1 shows the corresponding dataset.

There are various algorithms available to create such trees that are able to classify all given instances correctly: ID3/5 [32, 33], C4.5 [34], and CART [35], to name but

[1]It should be noted that the concept of regression models can also be used for classification. The equivalent algorithm would be logistic regression.

a few. All have in common that they follow several steps in order to built-up a tree: partitioning the data on a given variable (i.e., splitting), cutting of branches of the tree to reduce its size (i.e., pruning), and finally selecting the optimal tree that fits data best.

***Discussion*** As already demonstrated in the small example above, decision trees can be visualized and provide directly a means of interpretation of decisions made (at least for small trees). Decision trees can be applied to all problems, where linear regression would not be suited: e.g., non-linear problems, and can be applied to both regression and classification tasks. On the downside, trees can get quite complex with increasing height. Moreover, decision trees tend to overfit the data. Overfitting means that the decision tree is able to classify training data with high accuracy, but then fails to classify new, unseen data appropriately; overfitted models do not generalize well. Here, pruning of trees helps to reduce overfitting.

### 2.1.2 Surrogate Models

Another approach to introduce explainability to AI algorithms and systems is by replacing complex parts of an AI algorithm or an AI system by simpler models, which are more transparent, and thus ideally transform a black boxed system into a white boxed one. By using surrogate models to introduce explainability, one gains the advantage that explainability now becomes independent to the underlying AI algorithms, and thus developers have more flexibility in selecting the best algorithm to solve a given problem. As seen in the previous sections, classical approaches to machine learning that allow interpretability are often limited and exclude current (deep) neural networks. The overall goal of the following two approaches, which will be presented, is to extract meaningful information from a black box model to produce human-friendly explanations.

#### LIME

***Basics*** Local interpretable model-agnostic explanations (LIME) approximate the underlying black box AI model (e.g., a support-vector machine, a random forest, or a neural network) as good as possible with an interpretable model (e.g., a decision tree) in order to provide human-understandable explanations. LIME is doing that by putting the inputs into relationship with the outputs by treating the original AI model as a black box (cf. Fig. 5).

The key aspect of the approach LIME is taking is that a surrogate model is approximated locally (i.e., explaining the behavior for a given input), and not globally (i.e., explaining the entire AI model in general). What does that mean? Let's take image classification as an example. How to explain that the AI algorithm chooses that the left picture in Fig. 6 shows a parrot? Instead of trying to replicate the entire AI model, LIME selects local regions (here: parts of an image), and modifies these (here: removing parts of the image) in order to observe how small changes in the input image affect the prediction. In this example, when most parts of the image except the head

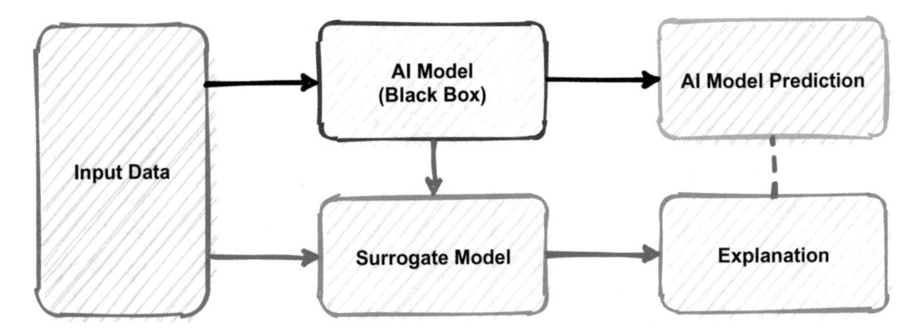


**Fig. 5** Abstract view of the pipeline of surrogate models such as LIME and SHAP, where the original AI model is considered a black box

of the parrot are removed, the probability that the hidden AI algorithms predicts the shown animal correctly is the highest (= 0.85).

***Discussion*** LIME is an intuitive approach to introduce explainability to any AI algorithm since it is agnostic to the underlying AI model. A potential disadvantage of LIME is certainly that only linear models are applied for model explanation, which might limit the overall applicability of LIME to explain complex AI models. It should further be noted that LIME is a rather new approach, and thus it cannot be said with high certainty that we can trust the explanations produced by LIME. For example, another group of researchers were able to produce any kind of explanation with LIME, meaning that the researchers were able to alter explanations in their interest, and thus questioning the trustworthiness of LIME itself [36]; the same applies for SHAP.

## SHAP

***Basics*** Shapley additive explanations (or SHAP [37]) is another class of an agnostic-model to produce explanations for arbitrary AI models. SHAP also relies on the underlying concept that input data is slightly modified in order to observe changes in the outcome (e.g., predictions) in order to create explanations. Specifically, SHAP relies on the concept of Shapley values, which are defined and applied excessively in cooperative game theory [38]. Without going into detail, it is sufficient to understand that we interpret each input feature as a player, and all players (features) work together towards a common goal (prediction). Shapley values try to distribute the individual payoff fairly, meaning that a player that invests more towards the common goal get a higher reward. Equivalently, a feature of the underlying AI model that has a stronger effect on the final prediction is more likely considered as an explanation than a feature that has hardly any impact on the final prediction. Several implementations[2] of SHAP exist: Tree SHAP, Deep SHAP, and Kernel SHAP.

[2]https://github.com/slundberg/shap.

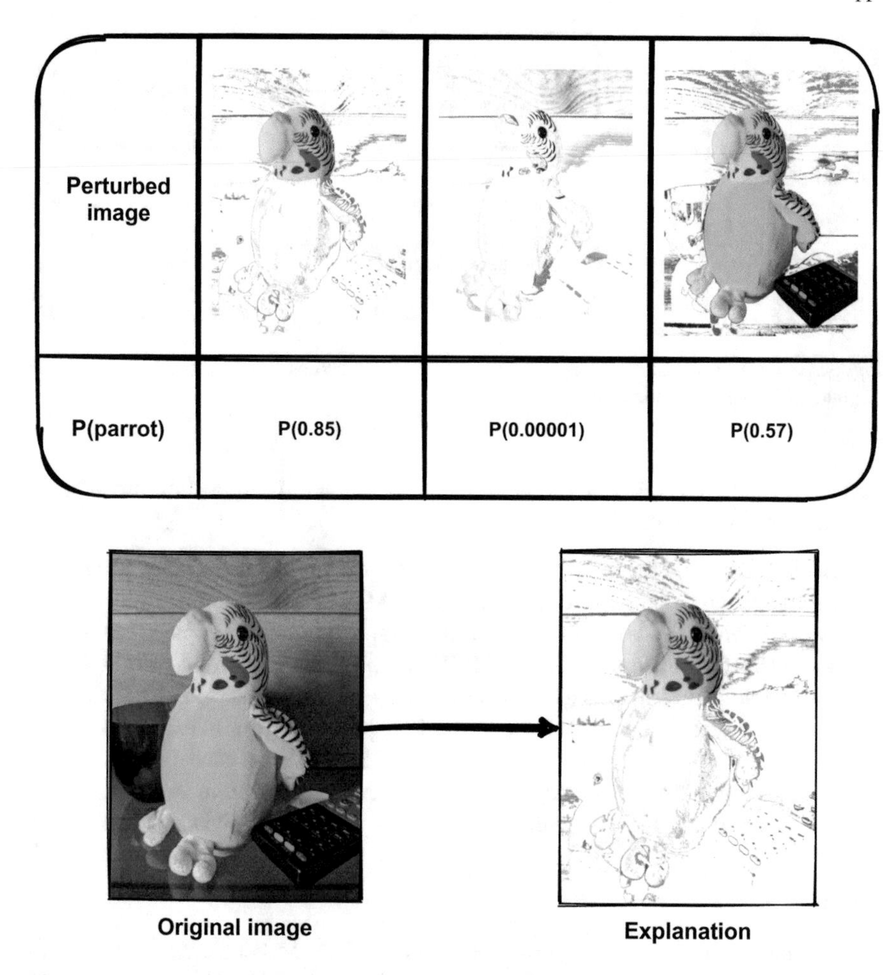


**Fig. 6** LIME: Transforming an image into interpretable components

***Discussion*** In contrast to LIME, the underlying concept of SHAP is sound and formally proven in game theory. Thus, SHAP gives a more robust impression with respect to the trustworthiness of explanations. Still, we have seen that an adversarial attack was affecting both, LIME and SHAP [36]. It will be interesting to see, what algorithms and frameworks will be developed on basis on LIME and SHAP.

## *2.2 Summary*

This section discussed ethical aspects of AI. In this context, two frameworks were presented that try to assess the trustworthiness and explainability of AI algorithms and systems: Ethical AI guidelines by the European Commission [14], and a traffic-light rating system proposed by VDE [26]. Besides these high-level frameworks, this section focused also on practical algorithms to maintain or increase the explainability of AI algorithms the necessity of introducing explainability to AI algorithms and systems. Both, classical models that allow human-understandable interpretation of AI results, e.g., linear regression and decision trees, as well as so-called surrogate models including LIME and SHAP, were briefly described and discussed. Where next? The level of explainability is dependent on the area of application (e.g., health domain versus recreational use), and the community that should be approached or interacts with an AI system (e.g., physician or a patient). So, different levels of explanations should be provided for different user groups, and thus systems should make use ideally of several technique to tackle the challenge of explainability. As proposed by [24], there will also be the need that both AI systems and users understand each other better. Acceptance of AI systems will greatly improve as soon as systems react to individual demands of users, and also make users aware if decisions made by AI need user examination. In the end, there is a contradiction to solve: reduce the "overtrust" of citizens in AI decisions, and at the same time, increase trustworthiness of AI solutions through improved transparency.

## 3 Advances in Algorithms

This section overviews some current research directions that will in future see more applications in real-world. We are leaving here the ethical aspects of AI and turn towards other emerging trends: automated machine learning [17], reinforcement learning [10], meta-learning [39, 40], and quantum AI [41].

### *3.1 Automated Machine Learning*

Both expert and non-expert users benefit to a great extent by automation processes. Since 80% of the daily work [42] of a data scientist consists of data cleaning, pre-processing, filtering, and preparation, automation of these tasks is essential. Automated machine learning (AutoML [17]), for example, is a research domain that provides a means to perform those fundamental tasks as well as feature engineering and hyper-parameter optimization in an automatic fashion. Having such functionality specifically allows non-experts to perform initial AI-driven workflows that yield reasonable results that can be used for an initial proof-of-concept solution and beyond. Thus, the overall goal is to automate common tasks of the data analytics pipeline (cf. Fig. 7). Currently, there are already a lot of tools available to support data scientists: auto-sklearn [43], autoWEKA [44], autoKeras [45], Google AutoML [46], or Amazon SageMaker Autopilot [47], to name but a few.

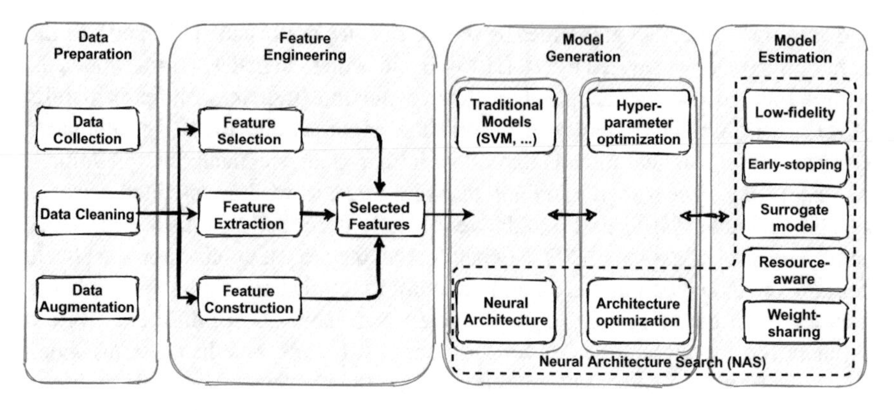


**Fig. 7** AutoML pipeline based on [17]

The set of models current AutoML libraries and services can choose from include classical machine learning algorithms such as SVMs, random forests, or k-means clustering. However, most of these libraries do not include yet neural networks and, specifically, deep learning algorithms; this is introduced by Network Architecture Search (NAS [19, 48, 49]), a new sub field of AutoML. NAS allows to design automatically neural network architectures such as convolutional neural networks (CNNs), and thus it is deemed the future of deep learning, since NAS lightens drastically today's workload of data scientists. It should be noted that AutoML and NAS free data scientists from time-consuming and cumbersome tasks, but this may come with a trade-off: accuracy. One could believe that hand-crafted AI pipelines and models might achieve a higher accuracy than automatically generated ones. In [50], the authors present EfficientNets, which describe a family of CNN models that, based on a NAS approach, proofed this presupposition wrong by achieving for selected problem statements a very high accuracy while producing smaller and faster models for inference.

## *3.2 Reinforcement Learning*

Although research about reinforcement learning (RL) dates back to the early 1950s [10], it just recently became of hot topic again in AI with prominent examples that made the news: AlphaGo [51] and its successor AlphaGo Zero [52]. Besides applying reinforcement learning to (video) games [53], there are more areas where RL is applied today: robotics [54], traffic light optimization [55].

But what is reinforcement learning? RL has its original in the evolutionary concept of trial-and-error, a method to solving problems of any kind. Specifically, small children are repeatedly trying out new things until they are successful. This involves a continuous learn process, where they not only learn on success, but moreover also

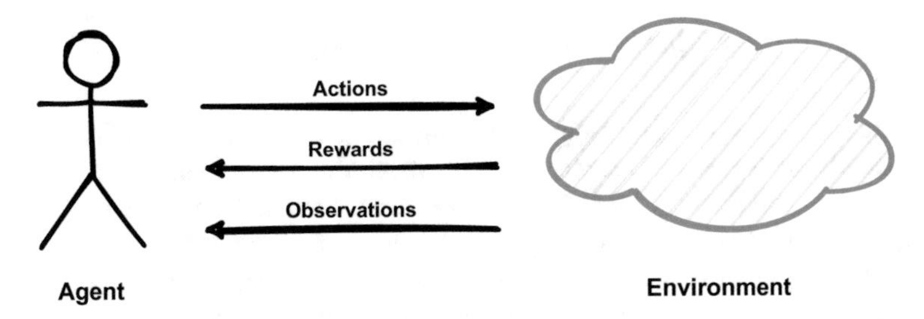


**Fig. 8** Basic workflow of reinforcement learning, where an agent interacts with its environment. An agent performs an action, and then receives feedback by the environment through observations and rewards. Based on these outcomes, the agent can adapt its next actions accordingly to increase the reward

on mistakes made. A child that slips on a wet surface while running will be more aware the next time and will run slower: the difference between pain and health.

The same basic concept underlies reinforcement learning. Here, the terminology of agents, actions, and rewards is introduced (cf. Fig. 8): an agent (e.g., a child) is taking an action (e.g., running on a wet surface), and expects some reward (e.g., being the first to cross an imaginary finish line). Based on the actual outcome, there is a feedback loop involved: a positive feedback when the action was successful (e.g., the child won), or a negative one when the action failed (e.g., the child fall down). So, the RL algorithm basically reacts with the environment and feeds gained knowledge back to the agent. Here, a typical goal is to maximize the rewards as quickly as possible. This could also mean that intermediate actions are not target-aimed, or at least make the impression that they are not beneficial (e.g., a chess player makes some unexpected moves before another figure is captured).

From an industry point of view, two dominant companies are providing toolkits for reinforcement learning: OpenAI Gym [56], and DeepMind Lab [57].

### *3.3 Meta-Learning*

Besides AutoML, Network Architecture Search, and reinforcement learning, another hot topic in AI is meta-learning. On top of the fundamental concept behind reinforcement learning ("trial and error"), meta-learning assumes that we learn through experience and actually do not start from scratch: we learn to learn more efficiently. Applying this concept in practice, one needs to take all data into account from previous AI pipeline (e.g.. input data, algorithm parameters and hyper-parameters as well as the evaluation results) in order to learn a better AI pipeline with higher accuracy. So, we learn from prior experience in a similar way as with reinforcement learning with the distinction that with meta-learning it would be possible to transfer a learned AI model to a new problem, e.g., applying an AI algorithm that plays Go for chess. Meta-learning is thus also a suitable tool to be integrated with AutoML.

### 3.4 *Quantum Artificial Intelligence (QAI)*

Machine learning as an interdisciplinary subject that has made outstanding achievements in different areas in recent years. However, due to its characteristic of data-intensive and compute-intensive demands, the pressure to find innovative approaches to machine learning is rising. A promising idea that is to exploit the potential of quantum computing in order to optimize classical machine learning algorithms. Quantum computation is appreciated in recent decades as it can employ unique characteristics of entanglement to speed up algorithms. Now, some quantum machine learning algorithms have been proposed and demonstrated including quantum support vector machine [58], supervised and unsupervised machine learning [59], quantum-enhanced machine learning [60], and distributed quantum learning [61].

Further, some schemes of quantum deep learning (QDL) have also been presented in research [62–64]. And quantum convolutional neural network based on the reverse-direction correspondence of the multi-scale entanglement re-normalization approach has been analyzed to realize efficient computation by exponentially reducing the number of parameters [63]. However, these schemes only focus on special models for particular tasks. The question is whether a more general and efficient scheme of QDL can be constructed to solve more tasks in full quantum process. A quantum deep learning scheme based on multi-qubit entanglement states, including computation and training of neural network in full quantum process, could result in an exponential speedup over classical algorithms.

## 4 Advances in Hardware: AI Accelerators

We have discussed so far ethical aspects of AI, and current trends in AI when it comes to recent progress made with respect to algorithms. This section shall complement the paper by giving a brief overview about existing and future AI-specific hardware including GPUs, FPGAs, ASICS, TPUs, and AI co-processors. Mathematics underlying most AI algorithms rely on classical linear algebra, probability theory and statistics. Specifically (deep) neural networks, auto-encoders, and deep belief networks rely on a vast amount of parallel executable vector and matrix operations, for which standard CPUs are not optimized for.

***GPUs*** A well-known AI accelerator are graphics processing units, GPUs. GPUs allow to process huge amounts of data simultaneously as required by current AI algorithms. A well-suited consumer GPU is the Nvidia GTX 1080Ti based on Nvidia's Pascal architecture. The GTX 1080Ti is equipped with 3,584 CUDA cores, and it is still today more than sufficient to perform typical workflows in AI. CUDA is the parallel computing model to implement applications on Nvidia GPUs by using language-specific extensions to map specific code blocks to the GPU instead of CPU. More recent iterations of Nvidia's GPU line are the RTX 2080 Ti (Turing architecture; 4,352 CUDA cores), or the RTX 8000 (4,608 CUDA cores; 576 Tensor cores). The difference between CUDA cores and Tensor cores is as follows: CUDA cores perform one single-precision operation per clock cycle. Tensor cores, by contrast, perform an entire $4 \times 4$ matrix operation per clock cycle. Although Tensor cores seem to perform more operations in a cycle than a CUDA core, the trade-off is loosing precision.

Server-class GPUs from Nvidia are based on its Volta architecture; the Tesla V100 has 5,120 CUDA cores and 640 Tensor cores. The next iteration, based on the Ampere architecture, was released on May 2020, and again yields a higher count of CUDA and Tensor cores. A recent alternative to Nvidia and CUDA was presented by AMD with its Radeon Open Compute (ROCm) platform, which offers to offload code to AMD GPUs in multi-node environments such as HPC systems.

***FPGAs, ASICs, and TPUs*** Aside from wide-spread GPUs, there exists also a niche for developing AI-specific hardware using FPGAs [65], ASICs [66], and TPUs [67]. FPGAs stand for field-programmable gate arrays, and this technology basically allows to reconfigure the underlying hardware based on a given purpose, e.g., to perform machine learning tasks more efficiently. Another advantage of FPGAs is that these cards are energy-efficient. Specifically in the domain of edge computing, where data analytics have to be performed in remote areas without access to power, FPGAs are first choice for inference (cf. Project Brainwave [68]). ASICs stand for application-specific integrated circuits. This kind of accelerator is able to achieve a higher performance than FPGAs due to its tailored hardware design while consuming even less power. On the downside, ASICS are not re-programmable and the initial production costs are high, so that this is usual not an option for individual usage. TPUs stand for tensor processing units, which are AI-specific ASICs developed by Google, which are available through Google's own cloud infrastructure [67]. The primary use of TPUs is to support TensorFlow applications.

There is some research regarding benchmarking GPUs, FPGAs, ASICs, and TPUs. We kindly refer the interested reader to these publications: [69–71].

## 5 Conclusions

We are currently living in an exciting era, where Artificial intelligence is no longer a buzzword lacking any relationship to society. AI has already transformed and advanced a multitude of areas of our everyday life and will continue to do so. With the growth of AI comes the need for more compute-intensive infrastructure to train models efficiently and generate inferences in near real-time, specifically when it comes to situations where human lives are at stack: e.g., autonomous driving or the entirety of the health sector. Aside from scientific and technological advances in AI, one challenge that arose more recently is the question how much trust we can and should put into AI systems. Several studies have shown that there is a concept of "overtrust" in AI systems, and thus citizens needs to be made aware about the techniques underlying AI systems and how to interpret predictions and further outcomes. As a consequence, the research field of explainable AI is specifically dealing with introducing transparency, trustworthiness and human-friendly explanations to AI systems.

In this paper, we gave an overview about the urgency to deal with ethical aspects of AI including its explainability. Here, we presented several approaches to introduce explanations to AI systems, and also highlighted some existing flaws. Furthermore, this paper presented some current trends in AI based on last year's hypes as identified by the annual Gartner study: automated machine learning, reinforcement learning, meta-learning, and quantum AI. Although not all fields mentioned are kind of "new", the have seen significant uptake in recent years. Since 80% of the work data scientist is centered around data preparation, there lies a huge potential in automating these steps. Furthermore, we will see tremendous advancements in quantum computing, and specifically quantum AI, over the next decades. All these advancements will not be feasible without significant progress in AI-specific hardware. In this paper, we already talked about GPUs, FPGA, ASICs, and so on. It will be interesting to see what the future holds.

**Acknowledgements** This work was supported by the research project CATALYST funded by the Ministry of Science, Research and the Arts of Baden-Württemberg, Germany (2016–2021).

## References

1. HLRS High Performance Computing Center Stuttgart—Annual Report (2020)
2. Farber, R., (ed.): AI-HPC is Happening Now (2017)
3. Usman, S., Mehmood, R., Katib, I.: Big Data and HPC Convergence: The Cutting Edge and Outlook (pp. 11–26). Springer, Cham (2017)
4. Brayford, D., Vallecorsa, S., Atanasov, A., Baruffa, F., Riviera, W.: Deploying AI frameworks on secure HPC systems with containers. In: 2019 IEEE High Performance Extreme Computing Conference (HPEC '19). IEEE (2019)
5. Jiang, Z., Wang, L., Xiong, X., Gao, W., Luo, C., Tang, F., Lan, C., Li, H., Zhan, J.: HPC AI500: the methodology. Roofline Performance Models, and Metrics for Benchmarking HPC AI Systems, Tools (2020)
6. Britt, K.A., Humble, T.S.: High-performance computing with quantum processing units. ACM J. Emerg. Technol. Comput. Syst. (JETC)
7. Grandinetti, L., Joubert, G.R., Michielsen, K.: Future Trends of HPC in a Disruptive Scenario, vol. 34 of Advances in Parallel Computing. IOS Press, Washington (2019)
8. Marr, B.: The Top 10 Artificial Intelligence Trends Everyone Should Be Watching In 2020. Forbes (2020)
9. Gartner.: Hype Cycle for Artificial Intelligence (2019)
10. Kaelbling, L.P., Littman, M.L., Moore, A.W.: Reinforcement learning: a survey. J. Artif. Intell. Res. **4**, 237–285 (1996)
11. Arulkumaran, K., Deisenroth, M.P., Brundagem, M., Bharath, A.A.: Deep reinforcement learning: a brief survey. IEEE Signal Process. Mag. **34**(6), 26–38 (2017)
12. Goebel,, R., Chander, A., Holzinger, K., Lecue, F., Akata, Z., Stumpf, S., Kieseberg, P., Holzinger, A.: Explainable AI: The New 42, pp. 295–303. Springer, Cham (2018)
13. Hagendorff, T.: The Ethics of AI Ethics: An Evaluation of Guidelines (2020)
14. High-Level Expert Group on Artificial Intelligence.: Ethics guidelines for trustworthy AI (2019)
15. Li, W., Liewig, M.: A Survey of AI Accelerators for Edge Environment, pp. 35–44. Springer, Cham (2020)

16. Wang, X., Han, Y., Leung, V.C.M., Niyato, D., Yan, X., Chen, X.: Convergence of edge computing and deep learning: a comprehensive survey. IEEE Commun. Surv. Tutor. **22**(2), 869–904 (2020)
17. He, X., Zhao, K., Chu, X.: A Survey of the State-of-the-Art. AutoML (2019)
18. Nagarajah, T., Poravi., G.: A review on automated machine learning (AutoML) systems. In: IEEE 5th International Conference for Convergence in Technology (I2CT). IEEE (2019)
19. Zoph, B., Le, Q.V.: Neural Architecture Search with Reinforcement Learning (2016)
20. Fenn, J., Raskino, M.: Mastering the Hype Cycle: How to Choose the Right Innovation at the Right Time. Inc./Harvard Business School Press series. Harvard Business Press, Boston, Mass, Gartner (2008)
21. Dedehayir, O., Steinert, M.: The hype cycle model: A review and future directions. Technol. Forecast. Soc. Change **108**, 28–41 (2016)
22. Hoy, M.B.: Alexa, Siri, Cortana, and more: an introduction to voice assistants. Med. Ref. Serv. Quart. **37**(1), 81–88 (2018)
23. Li, H., Ota, K., Dong, M.: Learning IoT in edge: deep learning for the internet of things with edge computing. IEEE Netw. **32**(1), 96–101 (2018)
24. Wagner, A.R., Borenstein, J., Howard, A.: Overtrust in the robotic age. Commun. ACM **61**(9), 22–24 (2018)
25. Marcus, G., Davis, E.: Rebooting AI: Building Artificial Intelligence We Can Trust, 1st edn. Pantheon Books, New York (2019)
26. Krafft, T., Hauer, M., Fetic, L., Kaminski, A., Puntschuh, M., Otto, P., Hubig, C., Fleischer, T., Grünke, P., Hillerbrand, R., Hustedt, C., Hallensleben, S.: From Principles to Practice: An Interdisciplinary Framework to Operationalise AI Ethics (2020)
27. Pasquale, F.: The black box society: the secret algorithms that control money and information. In: Pasquale, F. (ed.) The Black Box Society. Harvard University Press, Cambridge, Massachusetts and London, England (2015)
28. Samek, W., Montavon, G., Vedaldi, A., Hansen, L.K., Müller, K.-R.: Explainable AI: interpreting, explaining and visualizing deep learning. In: Samek, W., Montavon, G., Vedaldi, A., Hansen, L.K., Müller, K.-R. (eds.), LNCS Sublibrary. SL 7, Artificial Intelligence, Vol. 11700. Springer, Cham (2019)
29. Molnar, C.: Interpretable Machine Learning. LULU COM, [Place of publication not identified] (2020)
30. Ziegel, E.R., Myers, R.: Classical and modern regression with applications. Technometrics **33**(2), 248 (1991)
31. Stein, B., Lettmann, T.: Webis lecture notes: Decision trees (ml:iii) (2020)
32. Quinlan, J.R.: Induction of decision trees. Mach. Learn. **1**(1), 81–106 (1986)
33. Utgoff, P.E.: ID5: an incremental ID3. In: Laird, J. (ed.) Proceedings of the Fifth International Conference on Machine Learning ... 1988 ... Ann Arbor, pp. 107–120. Morgan Kaufmann, San Mateo, California (1988)
34. Ruggieri, S.: Efficient C4.5 [classification algorithm]. IEEE Trans. Know. Data Eng. **14**(2), 438–444 (2002)
35. Rutkowski, L., Jaworski, M., Pietruczuk, L., Duda, P.: The CART decision tree for mining data streams. Inform. Sci. **266**, 1–15 (2014)
36. Slack, D., Hilgard, S., Jia, E., Singh, S., Lakkaraju, H.: Adversarial Attacks on Post hoc Explanation Methods, Fooling LIME and SHAP (2019)
37. Lundberg, S.M., Lee, S.-I.: A Unified Approach to Interpreting Model Predictions, pp. 4765–4774 (2017)
38. Kuhn, H.W., Tucker, A.W.: Contributions to the theory of games (AM-28). Ann. Math. Stud. **28** (1953)
39. Vilalta, R., Drissi, Y.: A Perspective View and Survey of Meta-Learning, vol. 2 of 18. Artificial intelligence review edition (2002)
40. Vanschoren, J.: Meta-Learning: A Survey (2018)
41. Shaikh, T.A., Ali, R.: Quantum computing in big data analytics: a survey. In: IEEE Staff (eds) 2016 IEEE International Conference on Computer and Information Technology (CIT). IEEE (2016)

42. Press, G.: Cleaning Big Data: Most Time-Consuming, Least Enjoyable Data Science Task, Survey Says. Forbes (2016)
43. Feurer, M., Eggensperger, K., Falkner, S., Lindauer, M., Hutter. F.: Auto-Sklearn 2.0: The Next Generation (2020)
44. Kotthoff, L., Thornton, C., Hoos, H.H., Hutter, F., Leyton-Brown, K.: Auto-WEKA 2.0: automatic model selection and hyperparameter optimization in WEKA. J. Mach. Learn. Res. **18**(1), 826–830 (2017)
45. Jin, H., Song, Q., Hu., X.: Auto-Keras: an efficient neural architecture search system. In: Proceedings of the 25th ACM SIGKDD International Conference on Knowledge Discovery and Data Mining. Association for Computing Machinery (2019)
46. Bisong, E.: Google AutoML: Cloud Vision. In: Building Machine Learning and Deep Learning Models on Google Cloud Platform, pp. 581–598. Apress, Berkeley, CA (2019)
47. Das, P., Ivkin, N., Bansal, T., Rouesnel, L., Gautier, P., Karnin, Z., Dirac, L., Ramakrishnan, L., Perunicic, A., Shcherbatyi, I., et al.: Amazon SageMaker Autopilot: a white box AutoML solution at scale. In: Proceedings of the Fourth International Workshop on Data Management for End-to-End Machine Learning, pp. 1–7 (2020)
48. Bender, G., Kindermans, P.-J., Zoph, B., Vasudevan, V., Le, Q.: Understanding and simplifying one-shot architecture search. Int. Conf. Mach. Learn. 550–559 (2018)
49. Elsken, T., Metzen, J.H., Hutter, F.: Neural architecture search: a survey (2018). arXiv preprint arXiv:1808.05377 [Add to Citavi project by ArXiv ID]
50. Tan, M., Le, Q.V.: EfficientNet: Rethinking Model Scaling for Convolutional Neural Networks (2019)
51. Silver, D., Huang, A., Maddison, C.J., Guez, A., Sifre, L., van den Driessche, G., Schrittwieser, J., Antonoglou, I., Panneershelvam, V., Lanctot, M., Dieleman, S., Grewe, D., Nham, J., Kalchbrenner, N., Sutskever, I., Lillicrap, T., Leach, M., Kavukcuoglu, K., Graepel, T., Hassabis, D.: Mastering the game of Go with deep neural networks and tree search. Nature **529**(7587), 484–489 (2016)
52. Silver, D., Hubert, T., Schrittwieser, J., Antonoglou, I., Lai, M., Guez, A., Lanctot, M., Sifre, L., Kumaran, D., Graepel, T., Lillicrap, T., Simonyan, K., Hassabis, D.: A general reinforcement learning algorithm that masters chess, shogi, and Go through self-play. Science **362**(6419), 1140–1144 (2018)
53. Mnih, V., Kavukcuoglu, K., Silver, D., Rusu, A.A., Veness, J., Bellemare, M.G., Graves, A., Riedmiller, M., Fidjeland, A.K., Ostrovski, G., Petersen, S., Beattie, C., Sadik, A., Antonoglou, I., King, H., Kumaran, D., Wierstra, D., Legg, S., Hassabis, D.: Human-level control through deep reinforcement learning. Nature **518**(7540), 529–533 (2015)
54. Kober, J., Bagnell, J.A., Peters, J.: Reinforcement learning in robotics: A survey. Int. J. Robot. Res. **32**(11), 1238–1274 (2013)
55. Arel, I., Liu, C., Urbanik, T., Kohls, A.G.: Reinforcement learning-based multi-agent system for network traffic signal control. IET Intell. Transp. Syst. **4**(2), 128 (2010)
56. Brockman, G., Cheung, V., Pettersson, L., Schneider, J., Tang, J., Zaremba, W.: OpenAI Gym. John Schulman (2016)
57. Beattie, C., Leibo, J.Z., Teplyashin, D., Wainwright, M., Küttler et al.: DeepMind Lab, Tom Ward (2016)
58. Cai, X.-D., Wu, D., Su, Z.-E., Chen, M.-C., Wang, X.-L., Li, L., Liu, N.-L., Lu, C.-Y., Pan, J.-W.: Entanglement-based machine learning on a quantum computer. Phys. Rev. Lett. **114**(11), 110504 (2015)
59. Lloyd, S., Mohseni, M., Rebentrost, P.: Quantum algorithms for supervised and unsupervised machine learning (2013)
60. Dunjko, V., Taylor, J.M., Briegel, H.J.: Quantum-enhanced machine learning. Phys. Rev. Lett. **117**(13), 130501 (2016)
61. S, Y.B., Zhou, L.: Distributed secure quantum machine learning. Sci. Bull. **62**(14), 1025–1029 (2017)
62. Kieferová, M., Wiebe, N.: Tomography and generative training with quantum Boltzmann machines, vol. 96 (2017)

63. Cong, I., Choi, S., Lukin, M.D.: Quantum convolutional neural networks. Nat. Phys. **15**(12), 1273–1278 (2019)
64. Steinbrecher, G.R., Olson, J.P., Englund, D., Carolan, J.: Quantum optical neural networks. NPJ Quant. Inf. **5**(1), 1–9 (2019)
65. Trimberger, S.M.: Field-Programmable Gate Array Technology. Springer Science & Business Media (2012)
66. Smith, M.J.S.: Application-Specific Integrated Circuits, vol. 7. Addison-Wesley Reading, MA (1997)
67. Jouppi, N.P., Young, C., Patil, N., Patterson, D., Agrawal, G., Bajwa, R., Bates, S., Bhatia, S., Boden,, N., Al Borchers et al.: In-datacenter performance analysis of a tensor processing unit. In: Proceedings of the 44th Annual International Symposium on Computer Architecture, pp. 1–12 (2017)
68. Fowers, J., Ovtcharov, K., Papamichael, M.K., Massengill, T., Liu, M., Lo, D., Alkalay, S., Haselman, M., Adams, L., Ghandi, M. et al.: Inside project brainwave's cloud-scale, real-time AI processor. IEEE Micro **39**(3), 20–28 (2019)
69. Wang,, Y.E. Wei, G.-Y., Brooks. D.: Benchmarking TPU, GPU, and CPU platforms for deep learning. arXiv preprint arXiv:1907.10701 (2019)
70. Wang, Y., Wang, Q., Shi, S., He, X., Tang, Z., Zhao, K., Chu, X.: Benchmarking the performance and power of AI accelerators for AI training (2019). arXiv preprint arXiv:1909.06842
71. Reuther, A., Michaleas, P., Jones, M., Gadepally, V., Samsi, S., Kepner, J.: Survey and benchmarking of machine learning accelerators (2019). arXiv preprint arXiv:1908.11348

# Synergetic Build-up of National Competence Centres All over Europe

**Bastian Koller and Natalie Lewandowski**

**Abstract** This chapter presents the rationale behind and the implementation strategy for the setup of National Competence Centres for HPC and associated technologies all over Europe. Furthermore, it will present how a national activity like this, can benefit from coordination and support activities on the European level and how all this covers the needed actions in Europe to boost the uptake and impact of HPC.

## 1 HPC and Exascale in Europe

High-Performance Computing (HPC), as a key tool for science and industry, has steadily grown over the past decades into a mature technology that supports many of Europe's most important sectors. In many parts of the European economy, including engineering, health, climate and energy, usage of computer aided design coupled to modelling and simulation continues to grow rapidly. The software applications used in these sectors drive innovation.

In many areas of academia, industry but also public administration, the use of iterative modelling and simulation—including data management, analysis and visualisation—is becoming more and more important. HPC alone has matured and become a tool being taken up by more and more domains. The combination of it with associated technologies such as High Performance Data Analytics (HPDA) and Artificial Intelligence (AI) provide the means to tackle not only large, complex problems but also to widen further the use and uptake of these technologies in academia, public administration and industry.

---

B. Koller (✉) · N. Lewandowski
High Performance Computing Center Stuttgart, Nobelstrasse 19, 70569 Stuttgart, Germany
e-mail: koller@hlrs.de

N. Lewandowski
e-mail: lewandowski@hlrs.de



M. M. Resch et al. (eds.), *Sustained Simulation Performance 2019 and 2020*,
https://doi.org/10.1007/978-3-030-68049-7_13

Within the last years and with the decision on and the build-up of the EuroHPC Joint Undertaking,[1] the evolution of existing and establishment of new Centres of Excellence in HPC Applications was fostered (e.g. in the domain of Engineering,[2] Climate[3] or Biomedicine[4]).

Furthermore, a variety of ongoing research and development activities (hardware and software, e.g. the European Processor Initiative—EPI[5]) in the domain of HPC, HPDA and AI are continued and new projects, focused on Exascale-oriented technologies (also hardware and software, e.g. new CFD developments) are started. HPC centres worldwide are collaborating with their user communities in different disciplines and application-areas, and offer well-defined centre-specific mechanisms to support their user needs. At the same time, as we move towards Exascale, we see global developments which will lead to the availability of much more powerful systems in the coming years to cope with the ever-growing demand for computing resources. Nonetheless many of these things happen within the frame of EuroHPC, on the European level, but the corresponding national activities happen on a quite different level of maturity.

Thus, the best baseline for a European HPC Ecosystem would be a similar level of maturity of all the different nations, allowing a seamless synergetic collaboration between each of the entities and thus boosting the excellence and capabilities in that field in Europe. With that, the first phase of the activities towards National Competence Centres for HPC and associated technologies in Europe was born to start the establishing of the needed symbiotic cooperation of national centres. By that, stakeholders of HPC infrastructures, services and expertise in each nation are enabled to improve their service portfolio on base of the customers' needs and thus to help the end users with a maximum of efficiency in their respective domain.

## 2 Two Activities—One Common Goal

Looking at the current status of HPC—and associated technologies—related activities on a European level and supported within the frame of the EU H2020 program and the EuroHPC Joint Undertaking it is obvious that there is some awareness of results, potentials and achievements in the different nations. Figure 1 shows an outline of diverse activities on national and European level, which could be of interest for the National Competence Centres for HPC.

[1] EuroHPC—Leading the way in European Supercomputing; https://eurohpc-ju.europa.eu/.

[2] EXCELLERAT—The European Centre of Excellence for Engineering Applications; https://www.excellerat.eu/.

[3] ESiWACE—for future exascale weather and climate simulations; https://www.esiwace.eu/.

[4] e.g. BioExcel—Centre of Excellence for Computational Biomolecular Research; https://bioexcel.eu/.

[5] https://www.european-processor-initiative.eu/.

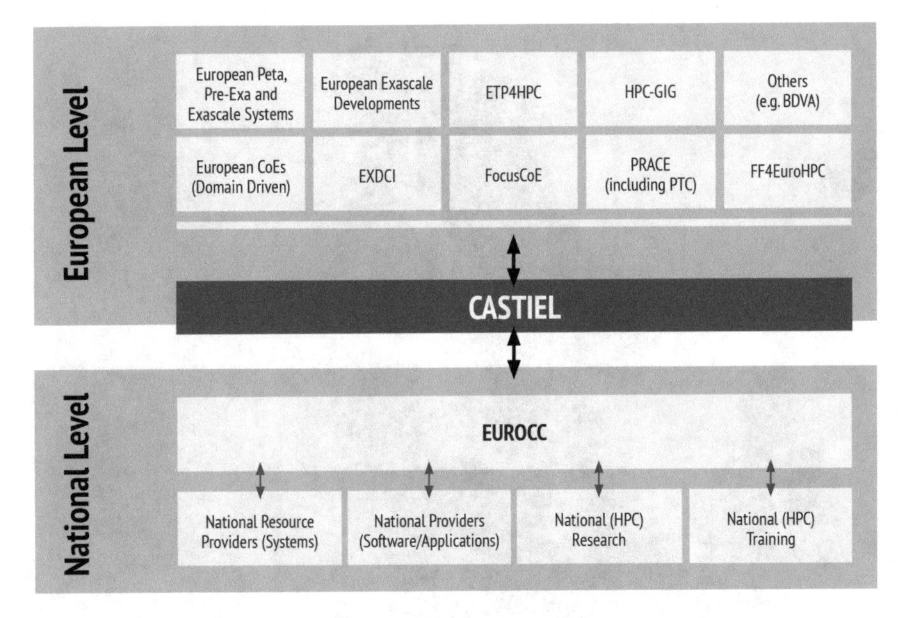


**Fig. 1** Closing the link between National and European activities

These activities are manifold and it is clear that those partners and countries involved in the activities have a high awareness on the aims and outcomes. Nonetheless not each entitiy providing competences in a respective field has the complete picture of what these activities are doing. Thus a hub approach, to collect the available capabilities would be an important asset to ensure awareness and thus allow for synergetic expansion of competences and expertise.

Finally, the baseline on national level is quite different, simply as evolution of competences for HPC, HPDA and AI are always a subject also of the national strategies, already available expertise and the degrees of investment.

It is clear that to achieve a similar level of competences all over Europe, there needs to be an activity to build up, on the one hand, a single reference point (hub) for (potential) HPC users and, on the other hand, to ensure that there is a cross-national exchange and symbiotic support, to allow quick and to-the-point knowledge transfers, also between the single reference points—a network of Competence Centres. This is realized by two project activities EuroCC[6] and CASTIEL[7] which are integrated in the ecosystem as presented in Fig. 1.

[6]EuroCC—National Competence Centres in the framework of EuroHPC; https://www.eurocc-project.eu/.

[7]CASTIEL—Coordination and Support for National Competence Centres on a European Level; https://www.castiel-project.eu/.

**Fig. 2** Geographical coverage of the EuroCC participating states

## *2.1 EuroCC—Implementing the National Competence Centres*

The EuroCC project consists of 33 nations (cf. Fig. 2), which, with support of their governments, will each set up a National Competence Centre (NCC) interconnecting with experts and competence providers in their country. Within EuroCC a framework to run the NCC is set up, including the identification and implementation of a fitting governance structure. Then, the potential services to be provided are meant to be identified and coordinated, all based on the identified needs of the stakeholders/users from public, academia and industry. Especially for the latter group, one part of EuroCC is to set up the mechanisms and tools to interact with industrial users und to provide them with tailored support—thus, to advent the use of HPC and associated technologies for their production activities.

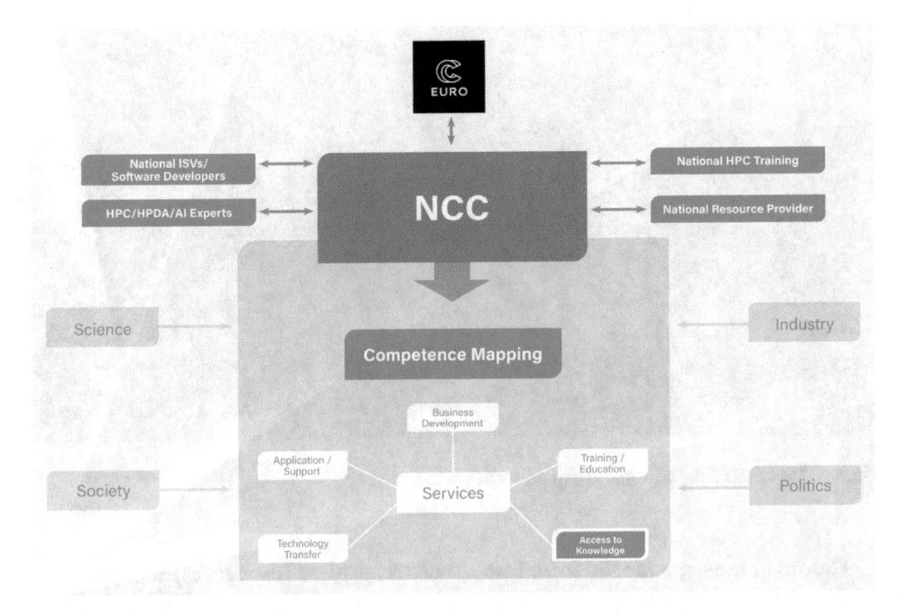


**Fig. 3** The high-level setup of the facets of a National Competence Centre

Figure 3 provides a high level diagram of the possible facets of a National Competence Centre. Some nations already cover a variety of the expected capabilities of the NCCs, whilst some only have competences available in parts of it. It is therefore part of the action to identify those available competences and to identify the needs in terms of existing gaps, to prepare a clear and to the point roadmap, which can also act as baseline for finding synergies with other NCCs.

Exactly this identification of synergies between the NCCs is a cumbersome activity, if each NCC is left alone with finding out, what the other NCCs are exactly planning and at which level of maturity they are. This is where CASTIEL enters the game, which is the activity supporting this information and knowledge flow and providing the means to foster guided and tailored interaction between the centres.

## 2.2 *CASTIEL—Coordination and Support on a European Level*

The CASTIEL Coordination and Support activity was designed to contribute to the success of the activities of the National Competence Centres as realized in EuroCC. Its main mission is to implement a framework of activities that will support the evolution of each single National Competence Centre and enable them step by step to get closer together in terms of capabilities and expertise. The greatest challenge hereby is the integration of the single nations into the overall strategy at the European level while preserving their autonomy and without interfering with their national strategy.

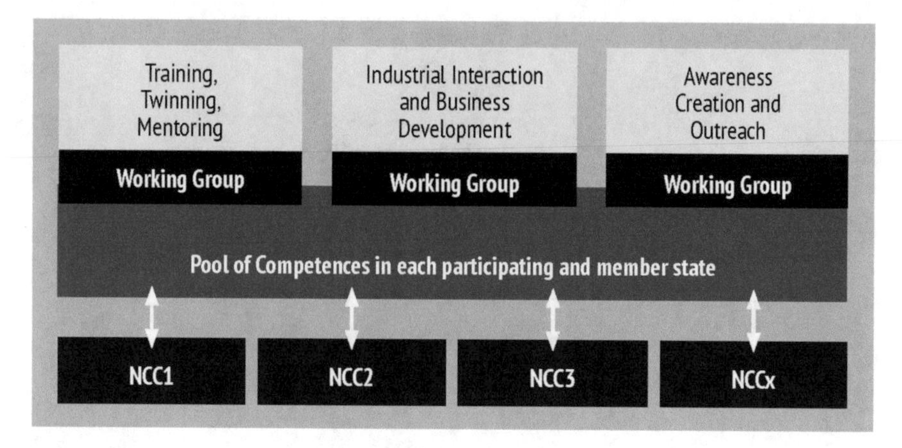


**Fig. 4** The CASTIEL working group concept

The main concept for the initial phase of the project (two years) will be to set up different working groups, which will elaborate on the best practices and needs of the National Competence Centres. These working groups are intended to use diverse communication mechanisms to ensure a strong interaction with and between the National Competence Centres on common topics, thus enabling an acceleration in evolution through the exchange of experience, knowledge and best practices. As a result of this, CASTIEL will also take over the implementation of exchange programs, workshops and other support activities.

As can be seen in Fig. 4, there are currently three major working groups foreseen on

- **Training, Twinning and Mentoring**. This working group promotes the consolidation of a complete and comprehensive European HPC Training programme supporting the NCCs requirements and needs. The focus will be to identify and catalogue the current state-of-the-art of training in HPC, HPDA and AI in the NCCs but also across Europe and at the same time to identify the gaps and needs. This will lead to an initial training proposal aimed to cover the identified training needs as well as the relevant actors but also into a comprehensive implementation of the twinning and mentoring programme, amongst others through workshop organisation.
- **Industrial Interaction and Business Development**. This working group supports those NCCs needing it, in their interaction with industry, and helps them to develop and improve their business models and industrial services. Here too, the identified competences of the NCCs will be the starting point. Via the working groups, CASTIEL coordinates the exchange and support, from one-to-one exchanges between two NCCs up to industrial training activities for a selected subset. Also, the support to transnational access to services is planned. This will allow a company to benefit from a symbiotic collaboration between NCCs, where the local NCC can extend its portfolio with the knowledge and the expertise of

another NCC. Furthermore, the access to other activities, such as the FF4EuroHPC project with open calls for business experiments or the PRACE Shape program, will be fostered within this working group.

- **Awareness Creation and Outreach**. This group supports the NCCs in their current dissemination of activities and to foster exchange of best practices between the NCCs and to create a EuroHPC NCC brand which represents the NCCs in a uniform design and integrated language on the European level. In addition to this, the baseline will be formed by the identified different sets of available competences and needs, taking into account the diverse target groups from industry, academia, and public. In collaboration with the other working groups, best practices are collected and a joint NCC knowledge base is implemented, including the highlighting of the NCCs added value for the respective target groups. This will also lead to a set of common dissemination and communication guidelines (regarding means, frequency, language, look and feel, and style). The means to be used for this can be participation at relevant conferences and activities of linked projects, distributing press releases and establishing media relations, newsletters and social media activities (e.g. the joint Twitter channel).

In the starting phase of the activities, each nation will be asked to identify their representatives for the respective working areas. They will create precisely the base for the pool of competences (such as champions representing their NCC), which is then used for further alignments, the evolution of the workplan and activities of the working groups.

## 3 Conclusions

EuroCC and CASTIEL have started their activities on the first of September 2020 and will run in a first phase for two years (until 31.08.2022). This phase will focus on the establishment of concepts and creation of the frame and implementation of the services and thus the National Competence Centres. Already now, EuroCC and CASTIEL have raised a vast interest all over Europe and beyond.[8,9] Also the social media presence is getting awareness and is continuously fed with highlights and information about the Competence Centres' evolutions (LinkedIn: EuroCC, Castiel-project; Twitter: @EuroCC'_project, @CASTIEL'_project).

With 33 nations being involved and following a common concept, there is a good opportunity for first success stories early within the runtime of the projects. The topics of Training, Twinning, Mentoring, Business Development, Industrial Interaction

[8]EuroCC and CASTIEL: two new projects to boost European HPC knowledge and opportunities—https://ec.europa.eu/digital-single-market/en/news/eurocc-and-castiel-two-new-projects-boost-european-hpc-knowledge-and-opportunities.

[9]GCS Coordinates Projects to Build Pan-European HPC Competency Network—https://www.hpcwire.com/off-the-wire/gcs-coordinates-projects-to-build-pan-european-hpc-competency-network.

and Awareness Creation have been identified as items of most common interest and are major topics to be supported by the Coordination and Support Action CASTIEL. The identification of available competences in the respective nations and their maintenance, will allow an accelerated bridging of gaps and provide the best basis for the evolution of the centres. Together with the creation of a knowledge base of related activities and the provided exchange possibilities, this will lead to a global European advancement. Thus, the NCCs will become an important part of the overall EuroHPC strategy, leading Europe into the area of Exascale Computing and beyond, and thereby fostering the uptake of HPC and associated technologies for industry, academia and public.

**Acknowledgements** The EUROCC project has received funding from the European High-Performance Computing Joint Undertaking (JU) under grant agreement No 951732. The JU receives support from the European Union's Horizon 2020 research and innovation programme and Germany, Bulgaria, Austria, Croatia, Cyprus, Czech Republic, Denmark, Estonia, Finland, Greece, Hungary, Ireland, Italy, Lithuania, Latvia, Poland, Portugal, Romania, Slovenia, Spain, Sweden, United Kingdom, France, Netherlands, Belgium, Luxembourg, Slovakia, Norway, Switzerland, Turkey, Republic of North Macedonia, Iceland, Montenegro. The CASTIEL project has received funding from the European High-Performance Computing Joint Undertaking (JU) under grant agreement No 951740. The JU receives support from the European Union's Horizon 2020 research and innovation programme and Germany, Italy, Spain, France, Belgium.

Printed in the United States
by Baker & Taylor Publisher Services